How do hormones and growth factors regulate animal growth in the developing embryo and after injury? What processes at the molecular level determine the growth patterns of different tissues? In this diverse synthesis of recent research, the regulation of growth in response to environmental and genetic stimuli is discussed at the level of the animal, tissues and cells. Contrasts are drawn between regulation in fetal and adult tissues, and in different tissues such as the CNS, bone and muscle. Functional chapters focus on the molecular links between mechanical tension and muscle growth, for example, while other chapters review the roles of specific first messenger molecules such as hormones and growth factors.

This state-of-the-art review will be of significant interest to graduate students and research scientists in the fields of animal growth, endocrinology and cell biology.

SOCIETY FOR EXPERIMENTAL BIOLOGY
SEMINAR SERIES: 60

MOLECULAR PHYSIOLOGY OF GROWTH

SOCIETY FOR EXPERIMENTAL BIOLOGY SEMINAR SERIES

A series of multi-author volumes developed from seminars held by the Society for Experimental Biology. Each volume serves not only as an introductory review of a specific topic, but also introduces the reader to experimental evidence to support the theories and principles discussed, and points the way to new research.

6. Neurones without impulses: their significance for vertebrate and invertebrate systems. *Edited by A. Roberts and B.M.H. Bush*
8. Stomatal physiology. *Edited by P.C.G. Jarvis and T.A. Mansfield*
16. Gills. *Edited by D.F. Houlihan, J.C. Rankin and T.J. Shuttleworth*
31. Plant canopies: their growth, form and function. *Edited by G. Russell, B. Marshall and P.G. Jarvis*
33. Neurohormones in invertebrates. *Edited by M. Thorndyke and G. Goldsworthy*
34. Acid toxicity and aquatic animals. *Edited by R. Morris, E.W. Taylor, D.J.A. Brown and J.A. Brown*
35. Division and segregation of organelles. *Edited by S.A. Boffey and D. Lloyd*
36. Biomechanics in evolution. *Edited by J.M.V. Rayner and R.J. Wootton*
37. Techniques in comparative respiratory physiology: An experimental approach. *Edited by C.R. Bridges and P.J. Butler*
38. Herbicides and plant metabolism. *Edited by A.D. Dodge*
39. Plants under stress. *Edited by H.C. Jones, T.J. Flowers and M.B. Jones*
40. *In situ* hybridisation: application to developmental biology and medicine. *Edited by N. Harris and D.G. Wilkinson*
41. Physiological strategies for gas exchange and metabolism. *Edited by A.J. Woakes, M.K. Grieshaber and C.R. Bridges*
42. Compartmentation of plant metabolism in non-photosynthesis tissues. *Edited by M.J. Emes*
43. Plant growth: interactions with nutrition and environment. *Edited by J.R. Porter and D.W. Lawlor*
44. Feeding and texture of foods. *Edited by J.F.V. Vincent and P.J. Lillford*
45. Endocytosis, exocytosis and vesicle traffic in plants. *Edited by G.R. Hawes, J.O.D. Coleman and D.E. Evans*
46. Calcium, oxygen radicals and cellular damage. *Edited by C.J. Duncan*
47. Fruit and seed production: aspects of development, environmental physiology and ecology. *Edited by C. Marshall and J. Grace*
48. Perspectives in plant cell recognition. *Edited by J.A. Callow and J.R. Green*
49. Inducible plant proteins: their biochemistry and molecular biology. *Edited by J.L. Wray*
50. Plant organelles: compartmentation of metabolism in photosynthetic cells. *Edited by A.K. Tobin*
51. Oxygen transport in biological systems: modelling of pathways from environment to cell. *Edited by S. Eggington and H.F. Ross*
52. New insights in vertebrate kidney function. *Edited by J.A. Brown, R.J. Balment and J.C. Rankin*
53. Post-translational modifications in plants. *Edited by N.H. Battey, H.G. Dickinson and A.M. Hetherington*
54. Biomechanics and cells. *Edited by F.J. Lyall and A.J. El Haj*
55. Molecular and cellular aspects of plant reproduction. *Edited by R. Scott and A.D. Stead*
56. Amino acids and their derivatives in plants. *Edited by R.M. Wallsgrove*
57. Toxicology of aquatic pollution: physiological, cellular and molecular approaches. *Edited by E. W. Taylor*
58. Gene expression and adaptation in aquatic organisms. *Edited by S. Ennion and G. Goldspink*
59. Animals and temperature: phenotypic and evolutionary adaptation. *Edited by I. A. Johnston and A. F. Bennett*

MOLECULAR PHYSIOLOGY OF GROWTH

Edited by

P. T. Loughna

Department of Veterinary Basic Sciences, The Royal Veterinary College, University of London, UK

J. M. Pell

Department of Cellular Physiology, Babraham Institute, Cambridge, UK

CAMBRIDGE UNIVERSITY PRESS
Cambridge, New York, Melbourne, Madrid, Cape Town, Singapore, São Paulo, Delhi

Cambridge University Press
The Edinburgh Building, Cambridge CB2 8RU, UK

Published in the United States of America by Cambridge University Press, New York

www.cambridge.org
Information on this title: www.cambridge.org/9780521114530

First published 1996
This digitally printed version 2009

A catalogue record for this publication is available from the British Library

Library of Congress Cataloguing in Publication data

Molecular physiology of growth/edited by P.T. Loughna and J.M. Pell.
p. cm. – (Society for Experimental Biology seminar series; 60)
Includes index.
ISBN 0 521 47110 9 (hardback)
1. Growth factors. 2. Cells–Growth–Molecular aspects.
I. Loughna, P.T. II. Pell, J.M. III. Series: Seminar series
(Society for Experimental Biology (Great Britain)); 60.
QP552.G76M65 1996
591.3–dc20 96–15182 CIP

ISBN 978-0-521-47110-7 hardback
ISBN 978-0-521-11453-0 paperback

Contents

Contributors

SKETH, J.
vett Research Institute, Greenburn Rd, Bucksburn, Aberdeen AB2
, Scotland, UK
LL, D.J.
vson Research Institute, St. Joseph's Health Centre, 268 Grosvenor
London, Ontario, Canada N6A 4VT
HNSTON, P.
nnah Research Institute, Ayr KA6 5HL, Scotland, UK
AN, R.U.
tion Research Laboratory for the Molecular Biology of Fetal Devel-
nent, Institute of Obstetrics and Gynaecology, Queen Charlotte's
Chelsea Hospital, RPMS, Goldhawk Road, London W6 0XG, UK
GAN, A.
partment of Clinical Chemistry, Wolfson Research Laboratories,
een Elizabeth Medical Centre, Edgbaston, Birmingham B15 2TH,

UGHNA, P.T.
partment of Veterinary Basic Sciences, The Royal Veterinary
llege, Royal College Street, London NW1 0TU, UK
ACKIE, E.J.
partment of Veterinary Basic Sciences, The Royal Veterinary
llege, Royal College Street, London NW1 0TU, UK
ARTIN, J.S.
partment of Medical Genetics, Glasgow University, Yorkhill,
asgow G3 8SJ, Scotland, UK
OORE, G.E.
tion Research Laboratory for the Molecular Biology of Fetal Devel-
ment, Institute of Obstetrics and Gynaecology, Queen Charlotte's
d Chelsea Hospital, RPMS, Goldhawk Road, London W6 0XG, UK
ELL, J.M.
epartment of Cellular Physiology, Babraham Institute, Cambridge
32 4AT, UK
AMSEY, S.
neumatology Research Unit, Addenbroke's Hospital, Cambridge CB2
2Q, UK
TICKLAND, N.C.
epartment of Veterinary Basic Sciences, The Royal Veterinary
ollege, Royal College Street, London NW1 0TU, UK
AUGHAN, J.I.
ction Research Laboratory for the Molecular Biology of Fetal Devel-
ment, Institute of Obstetrics and Gynaecology, Queen Charlotte's
d Chelsea Hospital, RPMS, Goldhawk Road, London W6 0XG, UK

Contributors

AKHURST, R.J.
Department of Medical Genetics, Glasgow Univ
Glasgow G3 8SJ, Scotland, UK
AKINSANYA, K.
Royal Postgraduate Medical School, Hammersmith H
W12 0HS, UK
ALI, Z.
Action Research Laboratory for the Molecular Biology
opment, Institute of Obstetrics and Gynaecology, Q
and Chelsea Hospital, RPMS, Goldhawk Road, Londo
BENNETT, P.R.
Action Research Laboratory for the Molecular Biology
opment, Institute of Obstetrics and Gynaecology, Qu
and Chelsea Hospital, RPMS, Goldhawk Road, Londo
BROWNSON, C.
School of Life Sciences, University of North London,
London N7 8DB, UK
CRILLY, P.J.
Hannah Research Institute, Ayr KA6 5HL, Scotland,
DICKSON, M.C.
Glaxo Research and Development Ltd, Greenford Ro
Middlesex UB6 0HE, UK
DWYER, C.M.
Department of Veterinary Basic Sciences, The Ro
College, Royal College Street, London NW1 0TU, UK
Present address: Department of Genetics and Behavi
Scottish Agricultural College, Edinburgh, Bush Estate,
0QE, Scotland, UK
FLINT, D.J.
Hannah Research Institute, Ayr KA6 5HL, Scotland, U
GLASSFORD, J.
Department of Cellular Physiology, Babraham Institu
CB4 2AT, UK

WYNICK, D.
Royal Postgraduate Medical School, Hammersmith Hospital, London W12 0HS, UK

D. J. FLINT, K. AKINSANYA, P. J. CRILLY,
P. JOHNSTON and D. WYNICK

The role of growth hormone in growth regulation

Introduction

The central role that growth hormone (GH) plays in growth and body composition has been documented extensively but in more recent times its role in a wide variety of functions, particularly in terms of reproduction and immune response, has begun to be explored more fully. In this chapter we describe the use of a model of GH deficiency involving passive immunization of rats with a specific antiserum to rat GH (anti-rGH) to investigate various aspects of development. The data provide evidence that GH plays a central role in development of both reproductive and immune functions *in vivo*, as well as demonstrating that GH plays a paradoxical role in stimulating adipocyte differentiation whilst enhancing lipid mobilization from mature adipocytes. Finally we describe the important autocrine/paracrine role that GH plays in the development of pituitary somatotrophs and its ability to sensitize the thyroid and ovary to the actions of TSH and the gonadotrophins, respectively.

Production and characterization of antibodies to rGH

Antisera to rGH were produced in sheep using a highly purified rGH preparation as immunogen. When examined in a radioimmunoassay, using ^{125}I-rGH, cross-reactivity with other pituitary hormones was very low and could be explained in terms of their contamination with rGH (Madon *et al.*, 1986). When the antiserum was assessed *in vitro* at the concentrations which could be achieved *in vivo*, it was capable of binding in excess of 1000 ng/ml of rGH. Since GH concentrations in the female rats used in this study are typically 100 ng/ml or lower, this indicated that the anti-rGH would effectively neutralize GH *in vivo* at the doses used.

A further important aspect of these studies was to be able to use recombinant bovine GH (bGH) replacement therapy in order to demonstrate that effects of the antiserum were due specifically to neutralization of GH. We were able to show that the antiserum to rGH cross-reacted extremely poorly with bGH *in vitro* and, when anti-rGH was administered *in vivo*, it failed to influence the half-life of injected bGH confirming its inability to interact with bGH. In preliminary studies we were able to show that treatment with anti-rGH for 24 h in young, rapidly growing rats, prevented body weight gain and dramatically reduced serum IGF-I concentrations, whilst concurrent treatment with bGH prevented these effects (Flint & Gardner, 1989).

Long-term treatment of neonatal rats with anti-rGH

Effects on growth

The fetal/neonatal rat is considered to be able to grow largely independently of GH with GH-dependent growth only developing around 3 weeks of age. We investigated this proposal by treatment with anti-rGH from 2 days of age, continuing treatment for either 3 weeks (short-term) or 8 weeks (long-term). Additional groups received bGH treatment concurrently with the anti-rGH. These studies clearly revealed that growth during the first 3 weeks of life was, at least in part, dependent upon GH since body weight gain was decreased around 20% (Flint & Gardner, 1993). If treatment with anti-rGH continued for the full 8 weeks then severe dwarfism occurred, clearly demonstrating the effectiveness of this approach (Fig. 1). Replacement therapy with bGH completely overcame the growth retardation. When short-term treatment ceased, the rats began to grow more rapidly, but they showed no evidence of catch-up growth, growing at a rate appropriate for their age rather than their weight. This contrasts with the catch-up growth shown by animals which are malnourished in early life. A variety of measures of long bone and muscle weights were determined and all were affected as anticipated based on the degree of dwarfism. Only one organ was demonstrated to be able to develop virtually independently of GH, the brain, results which support findings in neonatally hypophysectomized rats. Neonatally hypophysectomized rats do not live much beyond 30 d, and this was considered to be due to continued growth of the brain in the absence of appropriate cranial skeletal expansion (Glasscock *et al.*, 1990). If this is so, then it must be due to deficiencies other than GH since our dwarf rats, which are as severely stunted as the hypophysectomized rats, survived until at least 56 days of age when they were killed for tissue analysis. GH

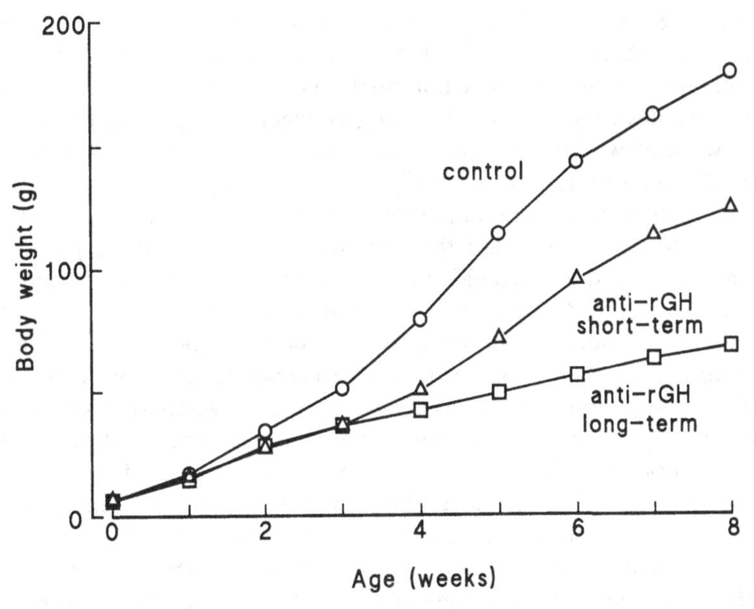

Fig. 1. Effect of anti-rGH treatment for 3 (short-term) or 8 (long-term) weeks on body weight in rats. (Adapted from Flint & Gardner, 1993.)

deficiency also led to a significant decrease in feed conversion efficiency in both short- and long-term treated groups (Flint & Gardner, 1993).

Effects on body composition

GH is known to increase lean : fat ratios by improving nitrogen retention and enhancing lipid mobilization in adipose tissue (for review, see Flint 1994). This was clearly shown to be the case in our studies since both the short-term and long-term GH deficient rats showed decreases in the proportion of lean tissue with increases in percentage body fat. In addition, those animals which received replacement therapy with bGH in sufficient quantities to achieve an equivalent body weight gain to that of control animals ended up with an increased lean : fat ratio compared with the controls, clearly demonstrating the ability of GH to decrease fat deposition.

In apparent contradiction to this role for GH, an almost entirely separate literature has evolved, describing the requirement for GH in the terminal differentiation of at least two preadipocyte cell lines derived from mice (Morikawa, Nixon & Green, 1982; Green,

Morikawa & Nixon, 1985). Because such studies have been described in cell lines, their relevance to the whole animal has been questioned. We therefore examined this potential role for GH *in vivo* using our model. By devising a technique using collagenase digestion of adipose tissue without washing the cells (since washing results in selective loss of small adipocytes) and with higher power magnification, we were able to identify, size and enumerate even the smallest unilocular adipocytes. Such studies revealed that, in two major internal fat pads, the parametrial and the perirenal, GH deficiency led to reduced numbers of mature adipocytes whilst concurrent GH therapy prevented this (Fig. 2). To our surprise, however, subcutaneous adipose tissue behaved quite differently with short-term GH deficiency resulting in a two-fold *increase* in the number of adipocytes present, whereas even severe dwarfism only resulted in a modest decrease in the number of adipocytes in this depot (Fig. 3). These results present direct evidence for the involvement of GH in adipocyte differentiation *in vivo* and, in addition, indicate that the sensitivity to GH is site-specific, which may allow conformational as well as compositional effects of GH to be explored.

Why should GH enhance fat mobilization whilst increasing adipocyte differentiation? If mature adipocytes are maintained in culture in a

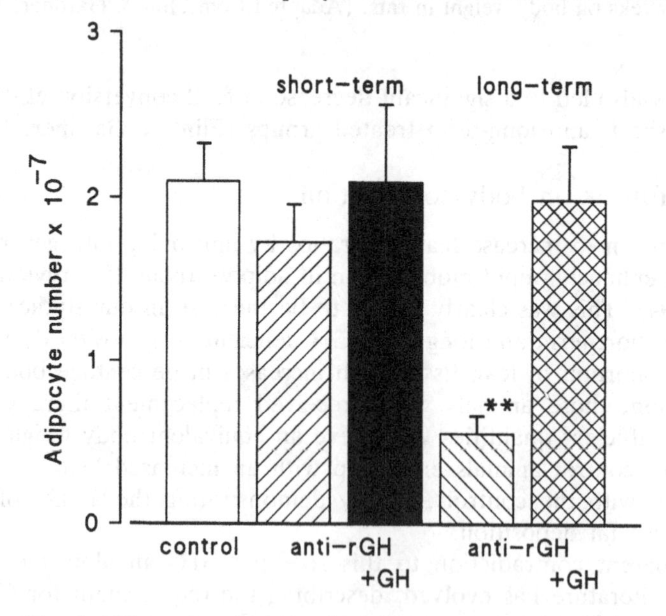

Fig. 2. Effect of anti-rGH and GH replacement therapy for 3 (short-term) or 8 (long-term) weeks on adipocyte numbers in parametrial adipose tissue. ** $p < 0.01$. (Adapted from Flint & Gardner, 1993.)

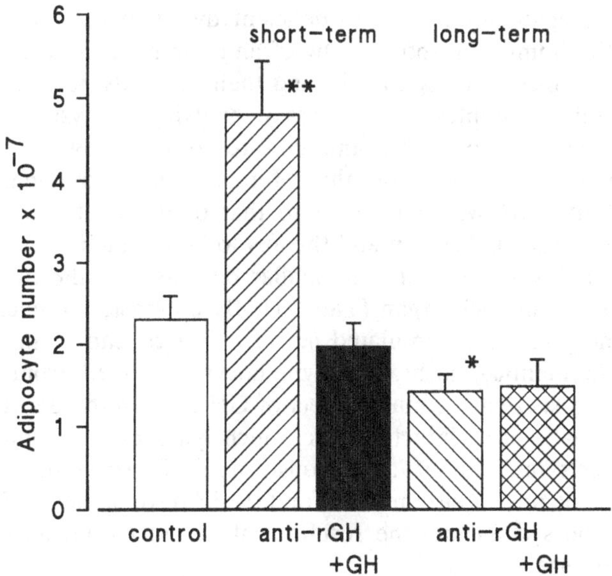

Fig. 3. Effect of anti-rGH and GH replacement therapy for 3 (short-term) or 8 (long-term) weeks on adipocyte numbers in subcutaneous adipose tissue. (Adapted from Flint & Gardner, 1993.) * $p < 0.05$, ** $p < 0.01$.

fashion which enhances lipid mobilization, such cells dedifferentiate and can subsequently undergo mitosis. Teleologically, it may seem beneficial to allow fat mobilization to occur under the influence of GH during periods of fasting, whilst GH prevents adipocyte dedifferentiation. Then, once refeeding occurs, lipid accumulation can occur rapidly, without requiring precursor adipocytes to undergo differentiation before lipid accumulation can occur.

GH and the immune system

In recent years there has been a growing interest in the interactions of the endocrine and immune systems, with evidence for the production of GH- and prolactin-like molecules by the immune system and feedback loops involving notably cytokines and the corticosteroid axis (Blalock, 1989). Earlier studies have suggested that GH can prevent atrophy of the thymus (Baroni, 1967) and various immune-deficiencies are evident in hypophysectomized rodents (see Blalock, 1989) or rats given anti-pituitary antibodies (Pierpaoli & Sorkin, 1968). The precise role of GH in development of the immune system and immune responsiveness has not been defined clearly in specific GH deficiency, however.

We therefore again used our GH-deficient dwarf to examine various aspects of the immune response. Dwarf and control rats received two injections of donkey red blood cells and their antibody responses were assessed in serum samples using an assay involving the lysis of donkey red blood cells *in vitro*. The antibody response in dwarf rats was severely compromised, whereas that of rats given recombinant bGH along with anti-rGH was equivalent to that of the control rats. GH-deficient dwarf rats had spleen and thymus weights which were severely reduced, and this was reflected in similar decreases in the number of lymphocytes within each organ (Table 1). By contrast, however, when these lymphocytes were stimulated *in vitro* with concanavalin A, they responded in identical fashion to lymphocytes derived from control rats. This suggests that the major effect of GH is upon the development of the epithelial cell architecture into which lymphocytes home rather than to defects in lymphocyte responsiveness. These studies support the proposal of an important role for GH in development of organs of the immune system and the resultant ability to mount an effective immune response.

Effects upon pituitary development

Severe dwarfism after long-term treatment with anti-rGH resulted in a decreased pituitary weight, principally due to a decrease in the number of somatotrophs and lactotrophs. This was also clearly evident from the marked reduction in both GH and prolactin secretion from dispersed pituitary cells of dwarf rats either under basal or stimulated conditions, when compared with control rats or rats given anti-rGH plus bGH replacement therapy (Akinsanya *et al.*, 1992 and Fig. 4). The reduction in prolactin secretion was to be expected since the

Table 1. *Lymphocyte numbers in spleen and thymus in rats treated for 8 weeks with anti-rGH, anti-rGH plus recombinant bGH or control rats*

	Cell number × 10^{-8}	
	Spleen	Thymus
Control	2.2 ± 0.3	3.8 ± 1.9
anti-rGH	0.7 ± 0.1**	2.2 ± 0.5*
anti-rGH plus bGH	2.2 ± 0.1	4.4 ± 1.3

*$p < 0.05$, **$p < 0.01$ compared with control values. Values are means ± SEM.

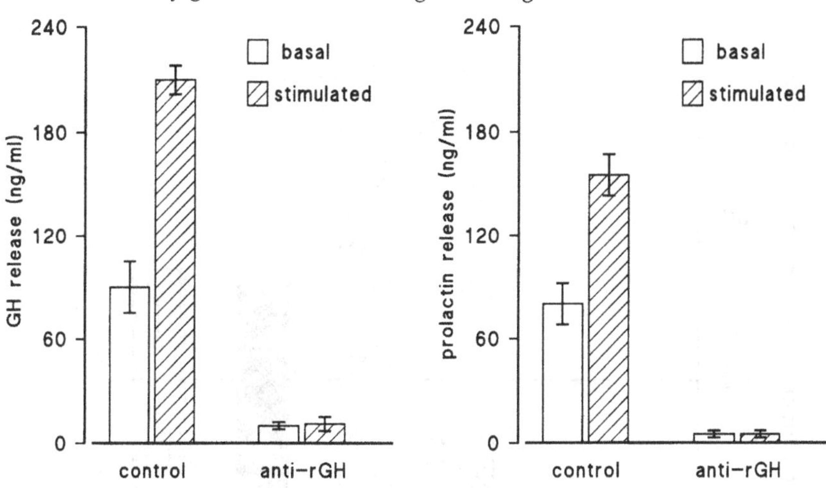

Fig. 4. Basal and growth hormone-releasing factor – (GRF-) stimulated GH release and basal and thyrotrophin releasing factor – (TRF-) stimulated prolactin release in control rats and long-term anti-rGH treated dwarf rats.

lactotrophs which secrete prolactin are derived from the somatotroph stem cell population. The precise mechanism by which anti-rGH leads to a reduction in the number of somatotrophs and lactotrophs, however, is not clear since it may have involved a cytotoxic effect in which the anti-rGH binds to rGH expressed on the surface of the somatotrophs leading to their destruction by complement-mediated cell lysis. This phenomenon does occur *in vitro* and it has been used in the absence of complement for FACS analysis of pituitary cells. Alternatively, the antibodies may have simply neutralized rGH and thereby prevented it from serving as a paracrine/autocrine factor for the somatotrophs themselves. We favour this latter explanation since, if the antiserum were simply cytotoxic, we would have expected exogenous bGH to overcome the effects of the anti-rGH on body growth as it does in hypophysectomized animals but we would not have expected it to normalize pituitary growth, as it did. Rather, we would have expected normal size rats with pituitaries deficient in somatotrophs and lactotrophs.

Pituitary contents of FSH, LH and TSH were relatively normal even in dwarf rats although to our initial surprise serum levels of all 3 hormones were significantly increased, particularly LH which was elevated 3-fold (Fig. 5).

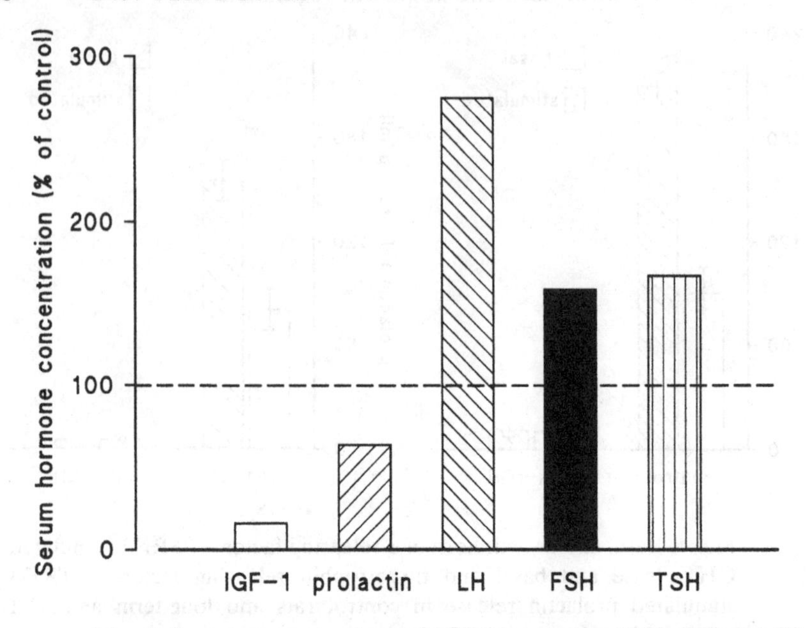

Fig. 5. Serum concentrations of IGF-I, prolactin, FSH, LH and TSH after long-term treatment with anti-rGH. Values are expressed as percentage of values in control rats. Serum GH concentrations could not be determined in the presence of high concentrations of anti-rGH in the serum samples and IGF-I was therefore used as an indicator of GH-deficiency.

A role for GH in sensitizing endocrine glands to their trophic hormones?

The elevated levels of several pituitary hormones could be explained in classic feedback terms since the development of the thyroid gland was impaired, with decreased T_3 and T_4 levels evident and the dwarfs were infertile even at 8 weeks of age (Flint *et al.*, 1992*b*) with decreased oestrogen and progesterone levels. Thus, one would anticipate increased release of pituitary FSH, LH and TSH in the face of decreased feedback from their respective target organs. This suggests, however, that even with elevated levels of the classical trophic hormones, the response in terms of end organ hormone production was severely diminished. That this was due to GH deficiency was demonstrated clearly by the fact that replacement therapy with recombinant bGH normalized serum progesterone, T_3, T_4, TSH, FSH and LH concentrations.

Mechanism of action of GH

Considerable debate still exists concerning the precise mechanisms of action of GH in regulating growth processes. On the one hand, most tissues possess GH receptors and a direct effect of GH is therefore plausible, whilst on the other hand, evidence exists for the effects of GH to be indirect, mediated via insulin-like growth factor-I (IGF-I) (Salmon & Daughaday, 1957). IGF-I does induce many of the changes produced by GH, for example in terms of increased long bone growth, body weight gain and nitrogen retention (see Cohick & Clemmons, 1993). In addition, treatment with GH leads to increased circulating concentrations of IGF-I, whilst treatment with anti-rGH induces a dramatic decrease in serum IGF-I concentrations. The effects of IGF-I are modulated by a family of IGF binding proteins (IGFBPs 1-6) and certain of these IGFBPs are influenced by GH treatment, notably IGFBP-3 upregulation and IGFBP-2 down regulation (see Baxter, 1991). It seems reasonable to imagine that these changes in IGF-I and its BPs in response to GH treatment might be involved in mediating the effects of GH. There are, however, a number of observations in a variety of systems which are at odds with this hypothesis. Perhaps the most clear cut evidence is the ability of GH to enhance fat mobilization from adipose tissue, since this can be demonstrated *in vitro* (i.e. in the absence of IGF-I) and adipocytes possess abundant GH receptors but apparently no IGF-I receptors (for review, see Flint & Vernon, 1993). In terms of growth, GH and IGF-I produce qualitatively different effects, with IGF-I producing particularly pronounced effects on kidney, thymus and spleen weights compared with GH. Effects of GH on growth could not be prevented using an antiserum to IGF-I (Spencer, Hodgkinson & Bass, 1991) whilst conversely an antiserum to rGH, given short-term (10 days) to growing rats, suppressed various parameters of growth without affecting circulating IGF-I concentrations (Palmer *et al.*, 1993).

Perhaps the most compelling case for an indirect effect of GH mediated via IGF-I relates to milk production since GH increases milk yield in the apparent absence of GH receptors on the mammary secretory epithelial cells (for review, see Bauman & Vernon, 1993). By contrast, there are IGF-I receptors on these cells and IGF-I has been shown, at least in the goat, to stimulate milk production (Prosser & Davis, 1992). The same authors, however, also provided evidence that the effects of GH and IGF-I were independent (Davis, Hodgkinson & Farr, 1991). Under appropriate conditions GH also stimulates milk secretion in rats but we were unable to mimic its effects with IGF-I, IGF-I and IGF-II precomplexed to IGFBP-3 or several analogues of IGF-I which are potentially more potent than IGF-I itself (Flint *et al.*, 1992*a*, 1994). We were also able to show that GH implants into the mammary gland were able to stimulate milk production only in the treated glands (i.e. not via a systemic route) (Flint & Gardner, 1994). Despite the fact that this effect of GH did not appear to be mediated via IGF-I, the time course of the effect suggested an indirect effect and herein may be the crux of this problem. Given the wide array of effects of long-term GH deficiency in growing rats, which exhibit multiple endocrine abnormalities, it is possible that, even in short-term GH deficiency, the sensitivity of numerous endocrine systems may be affected. In such circumstances there may be a plethora of 'mediators' of GH action. Such a situation does not, of course, rule out the possibility of additional, as yet unidentified, mediators of GH action.

References

Akinsanya, K., Wynick, D., Gardner, M.J., Flint, D.J. & Bloom, S.R. (1992). Function of cultured pituitary cells from rats treated neonatally with antibodies to rat growth hormone. *Journal of Endocrinology*, **132**, suppl. abst. 255.

Baroni, C. (1967). Thymus, peripheral lymphoid tissue and immunological responsiveness of the pituitary dwarf mouse. *Experimentia*, **23**, 282–3.

Bauman, D.E. & Vernon, R.G. (1993). Effects of exogenous bovine somatotrophin on lactation. *Annual Review of Nutrition*, **13**, 437–61.

Baxter, R.C. (1991). Insulin-like growth factor (IGF) binding proteins: the role of serum IFGBPs in regulating IGF availability. *Acta Paediatrica Scandinavia*, **372**, (Suppl.) 107–14.

Blalock, J.E. (1989). A molecular basis for bidirectional communication between the immune and neuroendocrine systems. *Physiological Reviews*, **69**, 1–32.

Cohick, W.S. & Clemmons, D.R. (1993). The insulin-like growth factors. *Annual Review of Physiology*, **55**, 131–53.

Davis, S.R., Hodgkinson, S.C. & Farr, B.C. (1991). The time-course of milk yield and hormonal responses following growth hormone injections in hourly-milked goats. *Proceedings of the New Zealand Society of Animal Production*, **51**, 239–44.

Flint, D.J. (1994). Immunomodulatory approaches for regulation of growth and body composition. *Animal Production*, **58**, 301–12.

Flint, D.J. & Gardner, M.J. (1989). Inhibition of neonatal rat growth and circulating concentrations of insulin-like growth factor-I using an antiserum to rat growth hormone. *Journal of Endocrinology*, **122**, 79–86.

Flint, D.J. & Gardner, M.J. (1993). Influence of growth hormone deficiency on growth and body composition in rats: site-specific effects upon adipose tissue development. *Journal of Endocrinology*, **137**, 203–11.

Flint, D.J. & Gardner, M.J. (1994). Evidence that growth hormone stimulates milk synthesis by direct action on the mammary gland and that prolactin exerts effects on milk secretion by maintenance of mammary deoxyribonucleic acid content and tight junction status. *Endocrinology*, **135**, 1119–24.

Flint, D.J., Gardner, M.J., Akinsanya, K. & Wynick, D. (1992*b*). Multiple endocrine deficiencies induced by an antiserum to growth hormone. *Journal of Endocrinology*, **132** suppl. abst. 60.

Flint, D.J., Tonner, D., Beattie, J. & Panton, D. (1992*a*). Investigation of the mechanism of action of growth hormone in stimulating lactation in the rat. *Journal of Endocrinology*, **134**, 377–83.

Flint, D.J., Tonner, E., Beattie, J. & Gardner, M.J. (1994). Several insulin-like growth factor-I analogues and complexes of insulin-like growth factors-I and -II with insulin-like growth factor binding protein-3 fail to mimic the effect of growth hormone upon lactation in the rat. *Journal of Endocrinology*, **140**, 211–16.

Flint, D.J. & Vernon, R.G. (1993). Hormones and adipose growth. In *The Endocrinology of Growth Development and Metabolism in Vertebrates*, ed. M.P. Schreibman, C.G. Scanes and P.K.T. Pang, pp. 469–494, San Diego: Academic Press.

Glasscock, G.F., Gelber, S.E., Lamson, G., McGee-Tekula, R. & Rosenfeld, R.G. (1990). Pituitary control of growth in the neonatal rat: effects of neonatal hypophysectomy on somatic and organ growth; serum insulin-like growth factors (IGF)-I and -II levels and expression of IGF binding proteins. *Endocrinology*, **127**, 1792–803.

Green, H., Morikawa, M. & Nixon, T. (1985). A dual effector theory of growth hormone action. *Differentiation*, **29**, 195–8.

Madon, R.J., Ensor, D.M., Knight, C.H. & Flint, D.J. (1986). Effects of an antiserum to rat growth hormone on lactation in the rat. *Journal of Endocrinology*, **111**, 117–23.

Morikawa, M., Nixon, T. & Green, H. (1982). Growth hormone and the adipose conversion of 3T3 cells. *Cell*, **29**, 783–9.

Palmer, R.M., Flint, D.J., McCrae, J.C., Fairhurst, F.E., Bruce, L.E., Mackie, S.C. & Lobley, G.E. (1993). Effects of growth hormone and an antiserum to rat growth hormone on serum IGF-I and muscle protein synthesis and accretion in the rat. *Journal of Endocrinology*, **139**, 395–401.

Pierpaoli, W. & Sorkin, E. (1968). Hormones and immunological capacity. I. Effects of heterologous anti-growth hormone (ASTH) antiserum on thymus and peripheral lymphatic tissue in mice. Induction of a wasting syndrome. *Journal of Immunology*, **101**, 1036–43.

Prosser, C.G. & Davis, S.R. (1992). Milking frequency alters the milk yield and mammary blood flow response to intra-mammary infusion of insulin-like growth factor-I in the goat. *Journal of Endocrinology*, **135**, 311–16.

Salmon, W.D., jr. & Daughaday, W.H. (1957). A hormonally controlled serum factor which stimulates sulphate incorporation by cartilage *in vitro*. *Journal of Laboratory and Clinical Medicine*, **49**, 825–36.

Spencer, G.S.G., Hodgkinson, S.C. & Bass, J.J. (1991). Passive immunization against insulin-like growth factor-I does not inhibit growth hormone-stimulated growth of dwarf rats. *Endocrinology*, **128**, 2103–9.

J.M. PELL and J. GLASSFORD

Insulin-like growth factor-I and its binding proteins: role in post-natal growth

Introduction

Insulin-like growth factor-I (IGF-I) is a potent growth factor, stimulating cell proliferation and differentiation, depending on its environment. It consists of a single 70 residue polypeptide chain which is structurally related to proinsulin (50% homology) and IGF-II (70% homology). In the original somatomedin hypothesis (for review, see Daughaday & Rotwein, 1989), it was originally identified as a sulphation factor which mediated the anabolic actions of growth hormone (GH). Mature IGF-I has a molecular weight of about 7500 D and consists of four major domains: B, C, A and D when named by analogy with the B, C and A regions of proinsulin. Its sequence is conserved across many diverse mammals and other vertebrates implying evolution from a common ancestral gene and signifying its importance in normal development.

Unlike classical endocrine hormones but in common with many peptide growth factors, IGF-I synthesis is ubiquitous, its mRNA being found in most tissues and its local autocrine/paracrine functions being well-established (D'Ercole, Stiles & Underwood, 1984). However, IGF-I is unusual amongst growth factors in its high blood concentrations which may exert endocrine actions. The liver is likely to be the major source of this circulating IGF-I as it has the greatest abundance of IGF-I mRNA (Murphy, Bell & Friesen, 1987), and its measured synthesis of IGF-I peptide can account for the known turnover of IGF-I in the circulation (Schwander *et al.*, 1983). In spite of its widespread distribution, very little IGF-I is present in the 'free' form, presumably due to its high potency; rather it is bound to one of six binding proteins which can regulate its bioactivity. IGF-I can bind to three receptors but has the greatest affinity for the type 1 IGF receptor, a ligand-stimulated tyrosine kinase which exhibits high homology to the insulin receptor, and is thought to mediate the mitogenic and differentiating properties of IGF-I and IGF-II (Czech, 1989). It will also bind, but with much lower affinity, to the insulin receptor and the IGF-II/mannose-6-phosphate receptor, whose precise function is uncertain.

It is clear from the above introduction that a complex and sophisticated system has evolved involving regulation of IGF-I action at various levels. This chapter will focus on key aspects of IGF-I gene expression in relation to growth control but will also briefly summarize the modulation of its subsequent actions by IGF binding proteins as they represent a key regulatory stage in the determination of IGF-I bioactivity.

The importance of IGF-I in growth regulation

Proliferative actions of IGF-I are easily exhibited *in vitro* in cell culture conditions but, even though modest increases in nitrogen balance and weight gain have been observed in intact rats (Tomas *et al.*, 1993*a*), anabolic actions of exogenous IGF-I have been difficult to demonstrate in normal healthy animals *in vivo*. However, much evidence exists to support the hypothesis that it is essential for normal growth regulation. For example, transgenic mice expressing GH only exhibit increased whole body growth rate after IGF-I gene expression is induced (Mathews, Norstedt & Palmiter, 1986). Growth rates in poodles (Eigenmann, 1985) and cattle (Hannon, Gronowski & Trenkle, 1991) can be related to IGF-I status. Administration of IGF-I to GH deficient and therefore IGF-I-deficient hypophysectomized rats (Schoenle *et al.*, 1982; Guler *et al.*, 1988) or Snell dwarf mice (Pell & Bates, 1992) can stimulate increased whole body weight gain. The absolute necessity of IGF-I for normal development has been demonstrated recently in two independent studies in which homozygous mice lacking a functional IGF-I gene ($IGF\text{-}I^{-/-}$) were generated by homologous recombination techniques. Even though the phenotype of the $IGF\text{-}I^{-/-}$ mice was comparable in both studies when compared with the wild-type mice, for example small size and underdeveloped skeletal muscle, those of Powell-Braxton *et al.* (1993) were more severely compromised than those reported by Liu *et al.* (1993) and Baker *et al.* (1993) and had a greater morbidity rate. It has been suggested (Powell-Braxton *et al.*, 1993) that this might be due to the position in the IGF-I gene in which the inactivating neomycin cassette was inserted (amino acid 15 versus 50); another explanation might be the degree of maternal rescue which is known to occur in some growth factor knock-out experiments.

In addition to the local tissue-derived IGF-I, circulating IGF-I may exert a powerful stimulation for growth control. Genetic selection of mice on the basis of high or low serum IGF-I concentrations resulted in a divergence in their growth rates (Blair *et al.*, 1989). Circulating

IGF-I concentrations are sensitive to GH (Zapf, Froesch & Humbel, 1981) and nutritional status (Prewitt *et al.*, 1982) and may have a role in mediating anabolic potential to peripheral tissues. In this regard, it is noteworthy that IGF-I can, in part, rescue the catabolism and weight loss which occurs during conditions such as diabetes, renal failure, gut resection and dexamethasone treatment (for example, see Tomas *et al.*, 1993*b*) which have been used as models for the wasting that accompanies in many acute clinical conditions. Thus, IGF-I can have anticatabolic properties and its clinical use may be appropriate in certain conditions.

Structure of the IGF-I gene

Even though mature IGF-I is a relatively small peptide, its gene is surprisingly large in mammals, comprising 80–100 kb of genomic DNA (Rotwein *et al.*, 1986; Shimatsu & Rotwein, 1987*a*; Sussenbach, Steenbergh & Holthuizen, 1992). Its overall structure is summarized in Fig. 1 for the human and rat genes. The gene has six exons, mature IGF-I being encoded by part of exon 3 and exon 4. Multiple mRNA species are derived from alternate splicing of exons upstream and downstream of exons 3 and 4, potentially yielding four prepro IGF-I species with a choice of two leader and two carboxy terminal sequences (for review see Lund, 1994). At the 5′ end of the gene, exons 1 and 2 may be

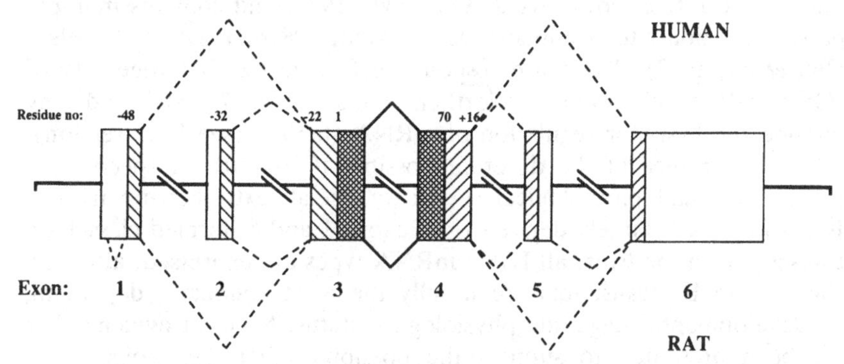

Fig. 1. Schematic structure of the human and rat IGF-I gene. Exons are shown as boxes and introns as the solid lines joining the boxes (both not to scale). Mature IGF-I is encoded by the cross-hatched regions of exons 3 and 4; other regions of translated mRNA are shown as diagonal shading. Alternate splicing of the leader and terminal exons is shown by dotted lines for human (above) and rat (below) IGF-I.

alternately spliced on to exon 3 to encode IGF-I leader sequences of 48 or 32 residues, respectively, of which 22 are encoded by exon 3. In the sheep, evidence for a further putative 5′ exon (termed exon 1W) has been found (Wong *et al.*, 1989; Dickson, Saunders & Gilmour, 1991) but to date practically negligible mRNA derived from this region has been detected (Pell, Saunders & Gilmour, 1993). Exon 4 also encodes 16 residues of the carboxy terminal or E peptides. The remainder of the E peptide is formed by different mechanisms in different species. In man, it is derived from alternate splicing of exon 4 to either exon 5 or exon 6 to encode E peptides with a total of 77 (termed Eb) or 35 (termed Ea) residues, respectively (Rotwein *et al.*, 1986). In rodents, the Ea peptide is formed in the same manner as in man but Eb is derived from the splicing together of exons 4, 5 and 6 resulting in a translational reading frame shift and encoding a 41 residue basic peptide lacking the *N*-glycosylation sites found in the Ea peptide (Roberts *et al.*, 1987; Shimatsu & Rotwein, 1987*a*). Exon 6-derived mRNA is the most abundant of the carboxy terminal species. Indeed, evidence for the existence of exon 5 has only recently been found in sheep; it is thought that the 3′ end of the ovine IGF-I gene is spliced in a similar manner to the human gene (J. Lyall and R.S. Gilmour, Babraham Institute, Cambridge, UK; personal communication). Exon 6 displays considerable homology between species, for example, encoding a similar number of residues. In this respect, it is noteworthy that exon 6-derived mRNA consists of about 6 kb of a 3′ untranslated region (3′UTR; Lund, Hoyt & Van Wyk, 1989) and contains multiple polyadenylation sites (Shimatsu & Rotwein, 1987*b*; Hoyt *et al.*, 1992; Hall *et al.*, 1992). These are responsible for the size heterogeneity of IGF-I mRNAs observed on Northern blots (0.9 to 7.5 kb), and may provide the basis for regulation of mRNA stability (see later section).

The occurrence of the different possible IGF-I mRNA species displays tissue and physiological specificity. In all extra-hepatic tissues, IGF-I mRNA is largely derived from exons 1 and 6, termed class 1-Ea transcripts. In the liver, all IGF-I mRNA types are expressed, although class 1 and Ea transcripts are usually the most abundant, depending on developmental stage and physiological status. No clear evidence has yet been presented to support the possibility that the choice of 5′ leader sequence could predetermine the splicing at the 3′ end of the gene.

Regulation of IGF-I synthesis and secretion

The complex structure of the IGF-I gene and the existence of the various IGF-I mRNA species may provide a means of differential and sensitive regulation of IGF-I synthesis and secretion which will be considered here.

IGF-I mRNA levels

Numerous studies conducted over the past decade have investigated the relationship between IGF-I mRNA abundance and tissue or whole body growth rate (the regulation of IGF-I gene expression has recently been reviewed by Zarrilli, Bruni & Riccio, 1994). Total hepatic IGF-I gene expression is regulated both by GH (Roberts *et al.*, 1986; Norstedt & Müller, 1987) and by dietary energy (Emler & Schalch, 1987; Strauss & Takemoto, 1991; Goldstein, Harp & Phillips, 1991) and protein content (VandeHaar *et al.*, 1991). Insulin may have a role in mediating the effects of changed feed intake, since it stimulates hepatic IGF-I gene expression (Tollet, Enberg & Mode, 1990; Houston & O'Neill, 1991; Boni-Schnetzler *et al.*, 1991).

To date, comparison of total exon 1 or 2 usage has been confined to hepatic tissue derived from ontogenic studies (Hoyt, Van Wyk & Lund, 1988; Adamo *et al.*, 1989) or from conditions of fasting, diabetes (Adamo *et al.*, 1991*a*) or GH deficiency (Lowe *et al.*, 1987; Butler *et al.*, 1994). In the non-interventional developmental investigations, differential levels of class 1 and 2 transcripts were observed with increases in exon 2-derived transcripts being observed at the time of onset of GH-dependent growth. The implied sensitivity to GH has been confirmed by studies in GH-treated hypophysectomized rats (Lowe *et al.*, 1987) and in normal sheep (Saunders *et al.*, 1991). However, Butler *et al.* (1994) could detect no significant differences in exon 1 and 2-derived mRNA in GH-treated *dw/dw* rats. Coordinate expression of class 1 and class 2 mRNA was reported for fasted and diabetic rats (Adamo *et al.*, 1991*a*). It is possible that the severe catabolism induced by fasting and diabetes or lifelong GH deficiency might induce responses which override normal physiological regulation. We therefore extended these findings to the investigation of levels exon 1 and 2-derived IGF-I mRNA in normal, intact animals in which the experimental variables were maintained within physiological limits.

Lambs were fed diets containing either low (L, 12%) or high (H, 20%) protein and fed either at a restricted level of intake (R, 3% of the animal's liveweight) or *ad libitum* (A, approximately 5% liveweight). Within these four dietary groups, GH (+; 0.1 mg/kg/day) or vehicle (−) was administered; the study started when the lambs were 10 weeks old and continued for a further 10 weeks (Pell *et al.*, 1993). Fig. 2 shows the percentage increase in hepatic class 1 and 2 transcripts over levels for the LR-group. Even though increases in exon

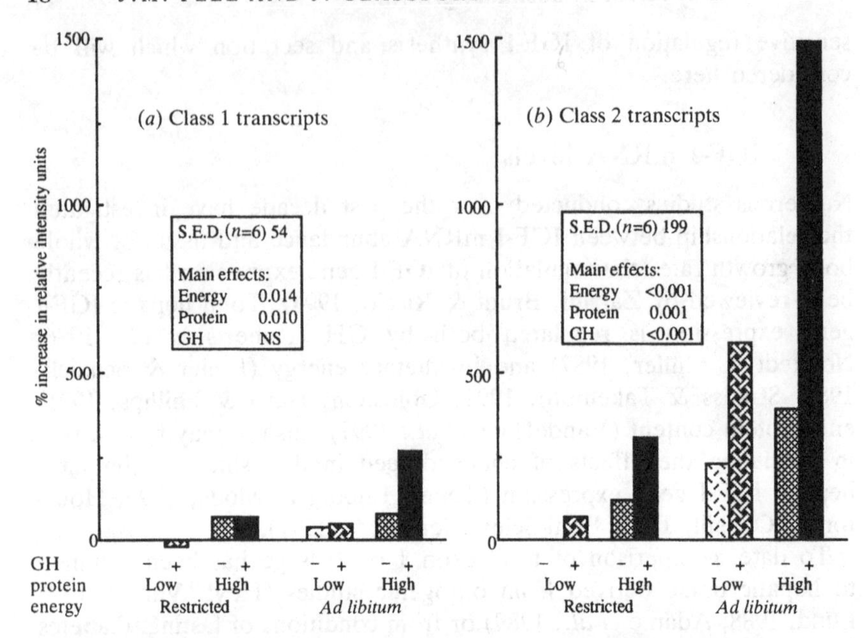

Fig. 2. Effect of GH status and dietary energy and protein on the percentage change in liver IGF-I mRNA on (*a*), class 1 transcripts and (*b*), class 2 transcripts. Correction was made for the number of U residues in each transcript. Percentage increases were calculated relative to low-protein, restricted-energy, saline-treated lambs which therefore have a value of zero.

1 and 2 mRNA levels were observed in response to improved nutritional status and to GH treatment, it is clear that the increases of exon 2-derived transcripts far exceeded those for exon 1. It is also apparent that the response of class 2 transcripts was greatest in GH versus non-GH treated animals. We decided to pursue this by selectively reducing GH synthesis in lambs by specific immunization against growth hormone-releasing hormone (GRF, Pell & Gilmour, 1993). Lambs were immunized at 6 weeks of age and boosted at monthly intervals. After 18 weeks, six of the GRF-immunized lambs were given a high dose of GH (0.25 mg/kg/day) for 1 week. Fig. 3 shows total hepatic IGF-I mRNA levels and the percent change in class 1 and class 2 transcripts compared with control-immunized lambs. GRF immunization induced a decrease in total hepatic IGF-I mRNA and a selective decrease in class 2 mRNA. Subsequent GH treatment restored IGF-I mRNA levels to values in excess of those for control lambs and also induced a differential increase in class 2 transcripts.

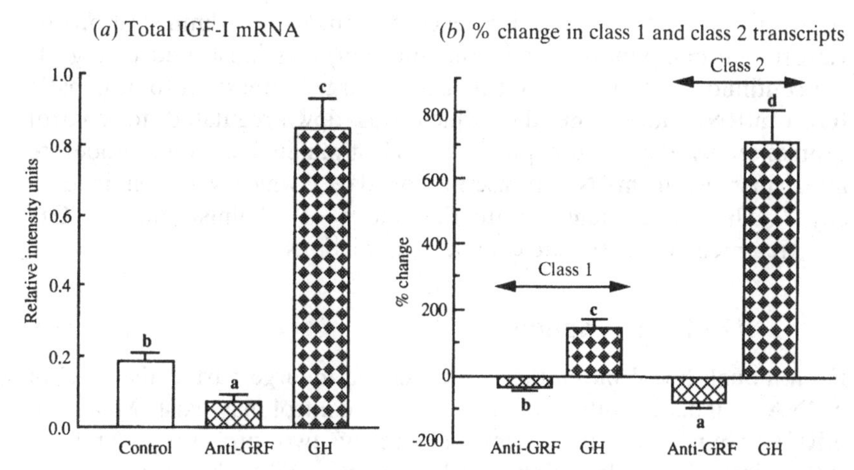

Fig. 3. Effects of immunization against GRF and subsequent GH treatment of (*a*), total liver IGF-I mRNA (data are means ± S.E.M.; a, b, c: means were statistically different $P<0.001$) and (*b*), the percentage change in hepatic class 1 and class 2 transcripts relative to mean levels in control lambs (data are means ± S.E.M.; statistical significance: a, b $P=0.013$; c, d $P<0.001$).

In both of these studies, hepatic IGF-I mRNA levels were correlated with circulating IGF-I concentrations. Skeletal muscle IGF-I mRNA abundance was determined and, as expected, this was almost exclusively derived from exon 1; muscle expression was at least 20-fold lower than in liver and was apparently unresponsive to changed GH and nutritional status. Studies in growing pigs have also confirmed the insensitivity of muscle IGF-I mRNA levels to increased GH concentrations in normal animals (Grant *et al.*, 1991). However, muscle growth rates do respond to GH and nutritional status, and these could only have been stimulated by autocrine/paracrine IGF-I if muscle IGF-I mRNA translatability was changed or if muscle IGF-I was rendered more active by binding protein modulation. It has therefore been proposed that hepatically synthesized endocrine IGF-I could have mediated any IGF-I induced changes in muscle growth. Moreover, it is likely that liver exon 2 was largely responsible for the increased circulating IGF-I concentration observed in the animals of high nutritional and GH status.

We have examined the temporal response to exogenous GH in dwarf rats, which exhibit a moderate GH deficiency (Charlton *et al.*, 1988), resulting in growth rates and IGF-I concentrations of about 50% of normal in our colony. Dwarf rats were treated with 70 to 500 μg GH per day for 1 or 7 days administered either by subcutaneous injection as two aliquots at 9am and 4pm each day (pulsatile administration) or via osmotic minipumps (continuous infusion). As illustrated in Fig. 4, the continuous infusion of GH induced a large increase in total hepatic IGF-I mRNA after one day which was down-regulated to control expression by day 7; the pulsatile GH stimulated a more moderate initial increase in mRNA transcripts on day 1 which was sustained to day 7. These data demonstrate that mode of administration of GH and treatment time exhibit complex interactions.

IGF-I transcription

Even though IGF-I biosynthesis is regulated in large part at the level of mRNA abundance, other potential sites of control may exist. Measured mRNA abundance is the net balance of two processes: synthesis (transcription) and degradation. Changes in mRNA abundance can be due to alterations in one or both of these processes. Clearly, mRNA levels alone can yield no information on such dynamics and represent

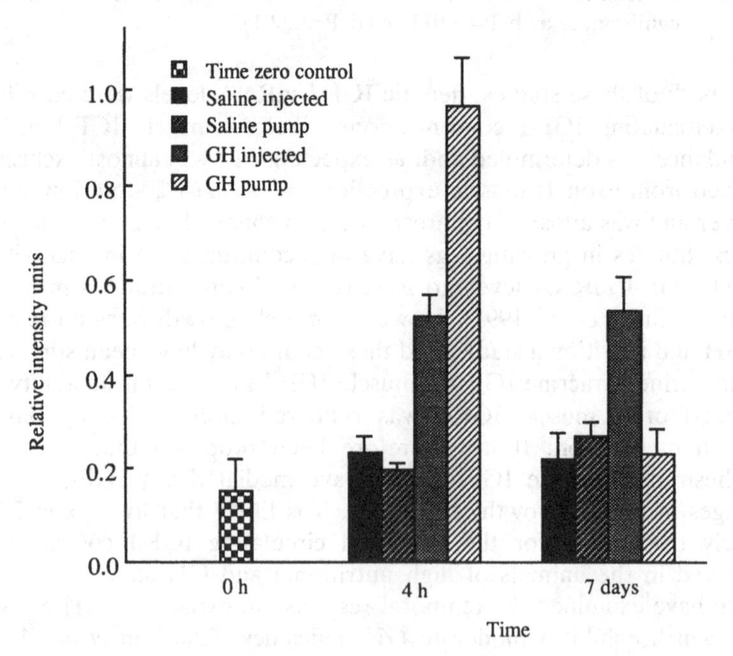

Fig. 4. Differential response of hepatic IGF-I mRNA levels to mode of administration and time of GH treatment in dwarf rats. Data are means ± S.E.M.; n=4 per group.

one time point only. In this section, transcriptional regulation of mRNA levels will be considered.

Studies using nuclear transcription elongation measurements and RNase protection assays employing intronic sequence for the labelled riboprobe (hence measuring the appearance of pre-mRNA) have demonstrated that changes in IGF-I mRNA levels can, in large part, be accounted for by transcriptional regulation of the IGF-I gene. For example, Hayden *et al.* (1994) have reported that acute fasting induces decreased levels of pre-mRNA. Acute treatment of hypophysectomized rats with a single dose of GH rapidly increases the appearance of nascent RNA (Bichell, Kikuchi & Rotwein, 1992). The early developmental (up to postnatal day 14) increase in hepatic IGF-I mRNA levels can be accounted for by increased transcription rate from exon 1 of the IGF-I gene (Kikuchi, Bichell & Rotwein, 1992). However, the same study demonstrates that, later in postnatal growth, mRNA derived from exon 2 increases at a greater rate than that derived from exon 1; the mechanisms underlying this are not yet defined. The factors regulating IGF-I gene transcription have been investigated by the identification of changes in chromatin structure (DNase 1 hypersensitive sites) and also possible regions of the gene involved in DNA–protein interactions. Kikuchi *et al.* (1992) have identified 15 DNase I hypersensitive sites in adult rat liver; during development, there was a progressive increase in some which coincided with the developmental increase in IGF-I mRNA levels. The same group (Bichell *et al.*, 1992) have also identified a single GH-responsive DNase I hypersensitive site, mapped to the second IGF-I intron (i.e. between exons 2 and 3). Protein–DNA interactions do occur in this region of the IGF-I gene but all were present in hypophysectomized rats and did not respond to acute GH treatment (Thomas *et al.*, 1995). At least 14 DNA–protein interactions have also been identified in the 1711 nucleotide region of promoter 1 and the 328 nucleotides of the 5′-untranslated region of exon 1 using *in vitro* and *in vivo* footprinting; none of these is GH sensitive (Thomas *et al.*, 1994). These data imply that simple changes in DNA–protein interaction cannot explain GH-induced IGF-I transcription; rather changes in tertiary DNA structure may occur involving additional protein-protein interactions. The position of a DNase I hypersensitive site between exons 2 and 3 indicates that distant modulators may interact with the proximal promoter regions of exons 1 and 2 to regulate transcription.

One important component of transcriptional regulation is the site of transcription initiation within the IGF-I leader exons and the regulatory elements in the associated promoters. In all mammals studied to date, transcription initiation within exon 1 occurs at several disperse and widely separated sites (usually 4 main sites) whereas transcripts derived from exon 2 are largely confined to a single cluster of sites (Adamo *et al.*, 1991*b*; Jansen *et al.*, 1991; Hall *et al.*, 1992; Shemer *et al.*, 1992; Pell *et al.*, 1993). In addition, the exon 1 promoter does not contain classical elements such as TATA or CCAAT or the GC-rich regions typical of housekeeping genes; putative TATA and CCAAT sequences have been suggested for regions upstream of the major exon 2 start site (Shemer *et al.*, 1992).

Transcripts derived from the exon 2 promoter are limited to a subset of tissues, unlike those for exon 1; these include liver, testis, lung, stomach and kidney (Shemer *et al.*, 1992). The exon 1 start site usage is, in part, tissue specific. Start site 3 is predominant in most tissues examined (testis, lung, stomach, heart, muscle) whereas equivalent usage of start sites 2 and 3 occurs in liver, kidney, pancreas and brain (Hall *et al.*, 1992; Shemer *et al.*, 1992). Exon 1 start site usage is coordinate in all tissues investigated during development except the kidney where start site 2 appears at weaning; interestingly this is also when transcripts derived from exon 2 are observed (Shemer *et al.*, 1992). Adamo *et al.* (1991*a*) have investigated start site usage in the liver of fasted and diabetic rats as well as during development; they have observed coordinate expression of all exon 1 start sites. Transcription derived from the single main site in exon 2 has been described above and was coordinate with exon 1 in the acute conditions but displayed a differential increase in the physiological development study at the onset of GH-dependent growth. Using an acute model of GH deficiency, Hall *et al.* (1992) and Bichell *et al.* (1992) reported coordinate start site usage within exon 1 and also between exons 1 and 2. However, under more chronic conditions, GH did not induce mRNA from all exon 1 start sites to the same extent (Hall *et al.*, 1992). Taken together, these data imply that short-term and long-term regulation of IGF-I mRNA species may be different.

Possible promoter regions responsible for exon 1 and 2 transcription have been investigated, usually using cell lines transfected with 5′ flanking fragments linked to a luciferase expression vector. For the rat, key regulatory regions of exon 1 are contained within the proximal 100–400 bp of 5′ flanking sequence; additional upstream negative elements were also identified (Lowe & Teasdale, 1992; Adamo *et al.*, 1993*a*). Promoter actively has been found for a wider region of 5′ flanking sequence for exon 1 of the human IGF-I gene (Kim, Lajara & Rotwein, 1991; Jansen *et al.*, 1992), although the two studies were not in complete agreement, possibly due to the use of different cell lines.

The putative exon 2 promoter has not been examined in as much detail. Adamo *et al.* (1993*b*) observed activity from −1.5 to −0.5 kb of flanking and 44 bp of exon 2 sequence using rat CHO cells but Hall *et al.* (1992) could not detect any exon 2 promoter activity using rat SK-N-MC cells.

Messenger RNA stability

Whilst it is thought that transcriptional regulation can, in large part, control IGF-I mRNA levels, mRNA degradation may be important in certain circumstances. Given the complexity of the IGF-I gene and its multiple mRNA products, it is not unlikely that some mRNAs may be less stable than others. Thus far, investigations on IGF-I mRNA degradation have focused on the function of the different sized transcripts rather than the role of specific leader sequences or start sites.

The 3′-UTR of IGF-I mRNA (derived from exon 6) is rich in A and U residues, which have been associated with a short mRNA half-life. The decay of the two major size species of IGF-I mRNA was therefore investigated *in vitro* and *in vivo* (Hepler, Van Wyk & Lund, 1990). *In vitro* (cell-free translation system), the larger species decayed more rapidly than did the small mRNA; *in vivo* (a single dose of GH to hypophysectomized rats), both mRNA species increased in abundance in parallel but the large species exhibited at least a 3-fold greater rate of decay than the small species. These data support the hypothesis that the long 3′-UTR may contribute to mRNA degradation. It might therefore be expected that stimuli which increase IGF-I transcription might preferentially induce the larger mRNA species since this could provide a rapid 'switch-on, switch off' mechanism. In support of this Johnson *et al.* (1991) have reported that GH treatment of primary rat hepatocytes induces the production of less stable IGF-I mRNA. Additionally, the liver which is largely responsible for sustained synthesis of circulating IGF-I, has only 30% of its total IGF-I mRNA in the 7.5–7.0 kg form whereas this species is responsible for nearly all extrahepatic IGF-I mRNA (Adamo *et al.*, 1993*b*). Low protein diets preferentially induce a decline in the abundance of the 7.5–7.0 mRNA (Thissen & Underwood, 1992). However, in spite of this circumstantial evidence, no rigorous examination has been performed to relate mRNA size and stability and anabolic stimuli. It is possible that the rat is an inappropriate model for the study of transcript size as its 3′ mRNA transcripts will always include transcripts derived from exon 6 either with or without exon 5, whereas in man and sheep exclusive use of exon 5 or 6 occurs.

RNA translatability

The extent to which the different lengths of IGF-I mRNA species affect their subsequent rates of translation has been investigated, particularly in relation to their multiple transcription initiation sites. Foyt *et al.* (1992) have determined the apparent *in vivo* translatability of IGF-I mRNAs with different 5′-UTRs by comparing their abundance in total and polysome populations. In control, hypophysectomized and GH-treated rats, exon 1-derived mRNAs were enriched in polysomal RNA in inverse proportion to the length of their 5′-UTR, as predicted by the scanning model of translation initiation. In contrast, exon 2 mRNAs, which have relatively short 5′-UTRs, were not enriched in polysomes suggesting that their translatability is regulated by additional factors or other properties associated with its mRNA. Thissen & Underwood (1992) have also examined the association of liver mRNA with polysomes based on transcript size; all major mRNA sizes were distributed in similar proportions in rats fed normal and low protein diets, indicating that all mRNA size classes were associated with IGF-I synthesis.

Physiological significance of IGF-I mRNA heterogeneity

Given the potency of IGF-I, its wide range of actions and its ubiquitous synthesis, it is perhaps not surprising that complex mechanisms have evolved which have the potential to differentially regulate its production. The relatively good conservation of the alternative amino and carboxy terminal sequences implies evolutionary and physiological significance. The liver exhibits the greatest degree of heterogeneity in terms of transcription and mRNA processing and is quantitatively the main site of IGF-I synthesis, being the major source of circulating IGF-I; it is consequently subject to the greatest degree of regulation. The overall conclusion of many studies is that hepatic IGF-I mRNA levels are very sensitive to the 'anabolic status' of an animal and mediate this by changes in circulating IGF-I concentrations. In addition, changes in hepatic mRNA derived from exon 2 are, in many models of normal growth regulation, more responsive than those derived from exon 1. This could merely imply that there are certain hepatic factors associated with exon 2 transcription or mRNA stability which are very responsive or, alternatively, that there is some significance to the mRNA and precursor peptide encoded by exon 2. To date, the role of the 5′ leader sequences in the predetermination of intracellular trafficking and subsequent fate of IGF-I has not been elucidated. Similarly, the function of the E peptides is not yet clear; the more abundant Ea peptide has two potential *N*-glycosylation sites which, due to a frame shift, are not present in the Eb peptides (Bach *et al.*, 1990). The pattern of expression of the Eb peptide parallels the time course for the increase in circulating IGF-I during development; this and its positive regulation by GH (Lowe *et al.*, 1988) have led Lund (1994) to suggest that the Eb domain could target IGF-I to the circulation. It is possible that *N*-terminal sequence differences can target IGF-I to specific intracellular locations, as does occur for other growth factors. Further, certain leader or E peptides may predispose IGF association with specific IGF binding proteins, although at least in the liver, cells responsible for IGF-I synthesis do not secrete all IGF binding proteins. The dominance of one species of IGF-I mRNA (derived from start site 3 of exon 1 and exon 6) in peripheral tissues, such as skeletal muscle indicates a less complex regulation than occurs in the liver which could be related to local autocrine/paracrine action. The distribution of mRNAs derived from exons 1 and 2 to all sizes of mRNA implies separate functions for the leader and exon 6 transcripts.

Role of IGF binding proteins

Only a small proportion of IGF-I exists in the free form, presumably due to its potency; instead a complex family of six IGFBPs, named IGFBP-1 to -6, have evolved to provide specific mechanisms for the regulation of IGF-I bioactivity. Their control of IGF-I is such that it would be inappropriate to consider IGF-I action in the absence of an awareness of IGFBP status. Nearly all cell types synthesize one or more of the six IGFBPs. They range in size from about 20 to 50 kD; IGFBP-3 can form an additional ternary complex with an acid-labile subunit and IGF-I which has a molecular weight of 150 kD and is responsible for the binding of over 90% of total IGF-I in the form of a circulating reservoir. The IGFBPs display considerable homology in specific regions of their amino and carboxy terminal sequences as well as containing specific structural domains. For example, most possess 18 cysteine residues which align and are probably involved in the conformation of structural motifs associated with IGF binding. IGFBP-1 and -2 also contain an arginine-glycine-aspartate (RGD) sequence which could enable them to bind to extracellular matrix-associated integrin receptors. IGFBP-3 and -4 are glycosylated, although the precise function for this is not yet clear. All IGFBPs will bind IGF-I and IGF-II, although with varying affinities; they do not bind insulin. The current literature on the physiology of the IGFBPs is vast and therefore their major putative functions will only be summarized here; instead, the following reviews provide thorough and excellent introductions to IGFBP action: Rechler (1994) and Jones & Clemmons (1995).

The precise role of the IGFBPs was, until recently, a matter of some speculation but all evidence to date suggests that they are involved in the regulation of IGF bioavailability. This includes the protection of cells from IGF-I action either via the type I or insulin receptors, the protection of IGF-I from degradation, the maintenance of a latent reservoir of bioavailable IGF-I (largely a function of the 150 kD IGFBP-3 ternary complex; Martin & Baxter, 1992), the shuttling and transport of IGF-I from the circulation (e.g. Bar *et al.*, 1990), the targeting of IGF-I in the cellular microenvironment (e.g. Arai *et al.*, 1994), and regulation of IGF activity (e.g. Elgin, Busby & Clemmons, 1987).

As well as controlling IGF-I activity, IGFBPs themselves are subject to regulation both at the level of synthesis rate and by posttranslational modifications (Clemmons, 1993; Jones & Clemmons, 1995). Usually, IGFBPs exhibit an affinity for IGF-I which is greater than that for the type I receptor and, in these conditions, IGFBPs will inhibit IGF-I action. However, IGFBPs may be modified by various mechanisms such as phosphorylation (Jones *et al.*, 1991), association with cell surfaces or the extracellular matrix (Conover *et al.*, 1993; Clemmons, 1993) or proteolytic modifications (Fowlkes *et al.*, 1994; Blat, Villaudy & Binoux, 1994). After each of these changes to the IGFBP protein, the affinity for IGF-I is reduced to a value equivalent or even less than that of IGF-I for the type 1 receptor; IGFBPs which were formerly inhibitory become potentiating and increase IGF-I activity.

Summary

IGF-I is a potent growth factor which can influence many aspects of cell proliferation and differentiation. Its gene structure is complex, allowing multiple tiers of regulation of IGF-I synthesis. In addition to the intricate nature of its expression, mature IGF-I is subject to further modification by association with a family of six IGF binding proteins which are themselves regulated at several levels. Thus a very sophisticated picture emerges in which IGF-I appears to be involved with many aspects of animal growth and development.

References

Adamo, M.L., Ben-Hur, H., LeRoith, D. & Roberts, C.T. Jr. (1991*a*). Regulation of start site usage in the leader exons of the rat insulin-like growth factor-I gene by development, fasting, and diabetes. *Molecular Endocrinology*, **5**, 1677–86.

Adamo, M.L., Ben-Hur, H., LeRoith, D. & Roberts, C.T. Jr. (1991*b*). Transcription initiation in the two leader exons of the rat IGF-I gene occurs from disperse versus localised sites. *Biochemical and Biophysical Research Communications*, **176**, 887–93.

Adamo, M.L., Lanau, F., Neuenschwander, S., Werner, H., LeRoith, D. & Roberts, C.T. Jr. (1993*a*). Distinct promoters in the rat insulin-like growth factor-I (IGF-I) gene are active in CHO cells. *Endocrinology*, **132**, 935–7.

Adamo, M.L., Lowe, W.L. Jr., LeRoith, D. & Roberts, C.T. Jr. (1989). Insulin-like growth factor I messenger ribonucleic acids with alternative 5′-untranslated regions are differentially expressed during development of the rat. *Endocrinology*, **124**, 2737–44.

Adamo, M.L., Neuschwander, S., LeRoith, D. & Roberts, C.T. Jr. (1993*b*) Structure, expression and regulation of the IGF-I gene. In *Current Directions in Insulin-Like Growth Factor Research*, ed. D. LeRoith and M.K. Raizada, pp. 1–11. New York: Plenum Press.

Arai, T., Arai, A., Busby, W.H. Jr. & Clemmons, D.R. (1994). Glycosaminoglycans inhibit degradation of insulin-like growth factor binding protein-5. *Endocrinology*, **135**, 2358–63.

Bach, M.A., Roberts, C.T. Jr., Smith, E.P. & LeRoith, D. (1990). Alternative splicing produces messenger RNAs encoding insulin-like growth factor-I prohormones *in vitro*. *Molecular Endocrinology*, **4**, 899–904.

Baker, J., Liu, J-P., Robertson, E.J. & Efstratiadis, A. (1993). Role of insulin-like growth factors in embryonic and postnatal growth. *Cell*, **75**, 73–82.

Bar, R.S., Boes, M., Clemmons, D.R., Busby, W.H., Sandra, A., Dake, B.L. & Booth, B.A. (1990). Insulin differentially alters transcapillary movement of intravascular IGFBP-1, IGFBP-2 and endothelial call IGF-binding proteins in the rat heart. *Endocrinology*, **125**, 1910–20.

Bichell, D.P., Kikuchi, K. & Rotwein, P. (1992). Growth hormone rapidly activates insulin-like growth factor I gene transcription *in vivo*. *Molecular Endocrinology*, **6**, 1899–908.

Blair, H.T., McCutcheon, S.N., Mackenzie, D.D.S., Gluckman, P.D., Ormsby, J.E. & Breier, B.N. (1989). Responses to divergent selection for plasma concentration of insulin-like growth factor-1 in mice. *Genetic Research (Cambridge)*, **53**, 187–91.

Blat, C., Villaudy, J. & Binoux, M. (1994). *In vivo* proteolysis of serum insulin-like growth factor (IGF) binding protein-3 results in increased availability of IGF to target cells. *Journal of Clinical Investigation*, **93**, 2286–90.

Boni-Schnetzler, M., Schmidt, C., Meier, P.J. & Froesch, E.R. (1991). Insulin regulates insulin-like growth factor I mRNA in rat hepatocytes. *American Journal of Physiology*, **260**, E846–51.

Butler, A.A., Ambler, G.R., Breier, B.H., LeRoith, D., Roberts, C.T. Jr. & Gluckman, P.D. (1994). Growth hormone (GH) and insulin-like growth factor-I (IGF-I) treatment of the GH-deficient dwarf rat: differential effects on IGF-I transcription start site expression in hepatic and extrahepatic tissues and lack of effect on type 1 IGF receptor mRNA expression. *Molecular and Cellular Endocrinology*, **101**, 321–30.

Charlton, H.M., Clark, R.G., Robinson, I.C.A.F., Goff, A.E.P., Cox, B.S., Bugnon, C. & Bloch, B.A. (1988). Growth hormone-deficient dwarfism in the rat: a new mutation. *Journal of Endocrinology*, **119**, 51–8.

Clemmons, D.R. (1993). Role of posttranslational modifications in modifying the biologic activity of insulin-like growth factor binding proteins. In *Current Directions in Insulin-Like Growth Factor Research*, ed. D. LeRoith and M.K. Raizada, pp. 245–253. New York: Plenum Press.

Conover, C.A., Clarkson, J.T., Durham, S.K. & Bale, L. (1993). Cellular actions of insulin-like growth factor binding protein-3. In *Current Directions in Insulin-Like Growth Factor Research*, ed. D. LeRoith and M.K. Raizada, pp. 255–266. New York: Plenum Press.

Czech, M. (1989). Signal transduction by the insulin-like growth factors. *Cell*, **59**, 235–8.

Daughaday, W.H. & Rotwein, P. (1989). Insulin-like growth factors I and II. Peptide, messenger ribonucleic acid and gene structures, serum, and tissue concentrations. *Endocrine Reviews*, **10**, 68–91.

D'Ercole, A.J., Stiles, A.D. & Underwood, L.E. (1984). Tissue concentrations of somatomedin C: further evidence for multiple sites of synthesis and paracrine or autocrine mechanisms of action. *Proceedings of the Society for Experimental Biology and Medicine*, **81**, 935–9.

Dickson, M.C., Saunders, J.C. & Gilmour, R.S. (1991). The ovine insulin-like growth factor-I gene: characterization, expression and identification of a putative promoter. *Journal of Molecular Endocrinology*, **6**, 17–31.

Eigenmann, J.E. (1985). Growth hormone and insulin-like growth factor in the dog: clinical and experimental observations. *Domestic Animal Endocrinology*, **2**, 1–16.

Elgin, G.R., Busby, W.H. Jr. & Clemmons, D.R. (1987). An insulin-like growth factor (IGF) binding protein enhances the biologic response to IGF-I. *Proceedings of the National Academy of Sciences, USA*, **84**, 3254–8.

Emler, C.A. & Schalch, D.S. (1987). Nutritionally-induced changes in hepatic insulin-like growth factor-I (IGF-I) gene expression in rats. *Endocrinology*, **120**, 832–4.

Fowlkes, J.L., Suzuki, K., Nagase, H. & Thailkill, K.M. (1994). Proteolysis of insulin-like growth factor binding protein-3 during rat pregnancy: a role for matrix metalloproteinases. *Endocrinology*, **135**, 2810–13.

Foyt, H.L., Lanau, F., Woloshak, M., LeRoith, D. & Roberts, C.T. Jr. (1992). Effect of growth hormone on levels of differentially processed insulin-like growth factor I mRNAs in total and polysomal mRNA populations. *Molecular Endocrinology*, **6**, 1881–8.

Goldstein, S., Harp, J.B. & Phillips, L.S. (1991). Nutrition and somatomedin XXII: molecular regulation of insulin-like growth factor-I during fasting and refeeding in rats. *Journal of Molecular Endocrinology*, **6**, 33–43.

Grant, A.L., Helferich, W.G., Kramer, S.A., Merkel, R.A. & Bergen, W.G. (1991). Administration of growth hormone to pigs alters the relative amount of insulin-like growth factor-I mRNA in liver and skeletal muscle. *Journal of Endocrinology*, **130**, 331–8.

Guler, H.P., Zapf, J., Scheiwiller, E. & Froesch, E.R. (1988). Recombinant human insulin-like growth factor I stimulates growth and has distinct effects on organ size in hypophysectomized rats. *Proceedings of the Society for Experimental Biology and Medicine*, **85**, 4889–93.

Hall, L.J., Kajimoto, Y., Bichell, D., Kim, S-W., James, P.L., Counts, D., Nixon, L.J., Tobin, G. & Rotwein, P. (1992). Functional analysis of the rat insulin-like growth factor I gene and identification of an IGF-I gene promoter. *DNA and Cell Biology*, **11**, 301–13.

Hannon, K., Gronowski, A. & Trenkle, A. (1991). Relationship of liver and skeletal muscle IGF-1 mRNA to plasma GH profile, production of IGF-1 by liver, plasma IGF-1 concentrations and growth rates in cattle. *Proceedings of the Society for Experimental Biology and Medicine*, **196**, 155–63.

Hayden, J.M., Marten, N.W., Burke, E.J. & Strauss, D.S. (1994). The effect of fasting on insulin-like growth factor-I nuclear transcript abundance in rat liver. *Endocrinology*, **134**, 760–8.

Hepler, J.E., Van Wyk, J.J. & Lund, P.K. (1990). Different half-lives of insulin-like growth factor I mRNAs that differ in length of 3′ untranslated sequence. *Endocrinology*, **127**, 1550–2.

Houston, B. & O'Neill, I.E. (1991). Insulin and growth hormone act synergistically to stimulate insulin-like growth factor-I production by cultural chicken hepatocytes. *Journal of Endocrinology*, **128**, 389–93.

Hoyt, E.C., Hepler, J.E., Van Wyk, J.J. & Lund, P.K. (1992). Structural characterization of exon 6 of the rat IGF-I gene. *DNA and Cell Biology*, **11**, 433–41.

Hoyt, E.C., Van Wyk, J.J. & Lund, P.K. (1988). Tissue and development specific regulation of a complex family of rat insulin-like growth factor I messenger ribonucleic acids. *Molecular Endocrinology*, **2**, 1077–86.

Jansen, E., Steenburgh, P.H., LeRoith, D. & Roberts, C.T. Jr. (1991). Identification of multiple transcription start sites in human insulin-like growth factor-I gene. *Molecular and Cellular Endocrinology*, **78**, 115–25.

Jansen, E., Steenbergh, P.H., van Schaik, F.M.A. & Sussenbach, J.S. (1992). The human IGF-I gene contains two cell type-specifically regulated promoters. *Biochemical and Biophysical Research Communications*, **187**, 1219–26.

Johnson, T.R., Rudin, S.D., Blossey, B.K., Ilan, J. & Ilan, J. (1991). Newly synthesized RNA: simultaneous measurement in intact cells of transcription rates and RNA stability of insulin-like growth factor I, actin, and albumin in growth hormone-stimulated hepatocytes. *Proceedings of the National Academy of Sciences, USA*, **88**, 5287–91.

Jones, J.I. & Clemmons, D.R. (1995). Insulin-like growth factors and their binding proteins: biological actions. *Endocrine Reviews*, **16**, 3–34.

Jones, J.I., D'Ercole, A.J., Camacho-Hubner, C. & Clemmons, D.R. (1991). Phosphorylation of insulin-like growth factor (IGF)-binding protein 1 in cell culture and *in vivo*: Effects on affinity for IGF-I. *Proceedings of the National Academy of Sciences, USA*, **88**, 7481–5.

Kikuchi, K., Bichell, D.P. & Rotwein, P. (1992). Chromatin changes accompany the developmental activation of insulin-like growth factor I gene transcription. *Journal of Biological Chemistry*, **267**, 21505–11.

Kim, S-W., Lajara, R. & Rotwein, P. (1991). Structure and function of a human insulin-like growth factor-I gene promoter. *Molecular Endocrinology*, **5**, 1964–72.

Liu, J-P., Baker, J., Perkins, A.S., Robertson, E.J. & Efstratiadis, A. (1993). Mice carrying null mutations of the genes encoding insulin-like growth factor I (*Igf-1*) and type 1 IGF receptor (*Igflr*). *Cell*, **75**, 59–72.

Lowe, W.T. Jr., Lasky, S.R., LeRoith, D. & Roberts, C.T. Jr. (1988). Distribution and regulation of rat insulin-like growth factor I messenger ribonucleic acids encoding alternative carboxyterminal E-peptides: evidence for differential processing and regulation in liver. *Molecular Endocrinology*, **2**, 528–35.

Lowe, W.L. Jr., Roberts, C.T. Jr., Lasky, S.R. & LeRoith, D. (1987). Differential expression of alternative 5′ untranslated regions in mRNAs encoding rat insulin-like growth factor I. *Proceedings of the National Academy of Sciences, USA*, **84**, 8946–50.

Lowe, W.L. Jr. & Teasdale, R.M. (1992). Characterization of a rat insulin-like growth factor I gene promoter. *Biochemical and Biophysical Research Communications*, **189**, 972–8.

Lund, P.K. (1994). Insulin-like growth factor I: molecular biology and relevance to tissue-specific expression and action. *Recent Progress in Hormone Research*, **49**, 125–48.

Lund, P.K., Hoyt, E.C. & Van Wyk, J.J. (1989). The size heterogeneity of rat insulin-like growth factor-I mRNAs is due primarily to differences in the length of the 3′ -untranslated sequence. *Molecular Endocrinology*, **3**, 2054–61.

Martin, J.L. & Baxter, R.C. (1992). Insulin-like growth factor binding protein-3: biochemistry and physiology. *Growth Regulation*, **2**, 88–99.

Mathews, L.S., Norstedt, G. & Palmiter, R.D. (1986). Regulation of insulin-like growth factor I gene expression by growth hormone. *Proceedings of the National Academy of Sciences, USA*, **83**, 9343–7.

Murphy, L.J., Bell, G.I. & Friesen, H.G. (1987). Tissue distribution of insulin-like growth factor I and II messenger ribonucleic acid in the adult rat. *Endocrinology*, **120**, 1279–82.

Norstedt, G. & Müller, C. (1987). Growth hormone induction of insulin-like growth factor I messenger RNA in primary cultures of rat liver cells. *Journal of Endocrinology*, **115**, 135–9.

Pell, J.M. & Bates, P.C. (1992). Differential actions of growth hormone and insulin-like growth factor-I on tissue protein metabolism in dwarf mice. *Endocrinology*, **130**, 1942–50.

Pell, J.M. & Gilmour, R.S. (1993). Differential regulation of IGF-I leader exon transcription. In *Current Directions in Insulin-Like Growth Factor Research*, ed. D. LeRoith and M.K. Raizada, pp. 13–23. New York: Plenum Press.

Pell, J.M., Saunders, J.C. & Gilmour, R.S. (1993). Differential regulation of transcription initiation from insulin-like growth factor-I (IGF-I) leader exons and of tissue IGF-I expression in response to changed growth hormone and nutritional status in sheep. *Endocrinology*, **132**, 1797–807.

Powell-Braxton, L., Hollingshead, P., Warburton, C., Dowd, M., Pitts-Meek, S., Dalton, D., Gillett, N. & Stewart, T.A. (1993). IGF-I is required for normal embryonic growth in mice. *Genes and Development*, **7**, 2609–717.

Prewitt, T.E., D'Ercole, A.J., Switzer, B.R. & Van Wyk, J.J. (1982). Relationship of serum immunoreactive somatomedin-C to dietary protein and energy in growing rats. *Journal of Nutrition*, **112**, 144–50.

Rechler, M.M. (1994). Insulin-like growth factor binding proteins. *Vitamins and Hormones*, **47**, 1–114.

Roberts, C.T. Jr., Brown, A.L., Graham, D.E., Seelig, S., Berry, S., Gabbey, K.H. & Rechler, M.M. (1986). Growth hormone regulates the abundance of insulin-like growth factor-I mRNA in adult rat liver. *Journal of Biological Chemistry*, **261**, 10025–8.

Roberts, C.T. Jr., Lasky, S.R., Lowe, W.L. & LeRoith, D. (1987). Rat IGF-I cDNAs contain multiple 5′ -untranslated regions. *Biochemical and Biophysical Research Communications*, **146**, 1154–8.

Rotwein, P., Pollock, K.M., Didier, D.K. & Krivi, K.K. (1986). Organization and sequence of the human insulin-like growth factor-I gene. *Journal of Biological Chemistry*, **261**, 4828–32.

Saunders, J.C., Dickson, M., Pell, J.M. & Gilmour, R.S. (1991). Expression of a growth hormone-responsive exon of the ovine insulin-like growth factor-I gene. *Journal of Molecular Endocrinology*, **7**, 233–40.

Schoenle, E., Zapf, J., Humbel, R.E. & Froesch, E.R. (1982). Insulin-like growth factor I stimulates growth in hypophysectomized rats. *Nature*, **269**, 252–3.

Schwander, J., Hauri, C., Zapf, J. & Froesch, E. (1983). Synthesis and secretion of insulin-like growth factor and its binding protein by the perfused liver: dependence on growth hormone status. *Endocrinology*, **113**, 297–305.

Shemer, J., Adamo, M.L., Roberts, C.T. Jr. & LeRoith, D. (1992). Tissue-specific transcription start site usage in the leader exons of the rat insulin-like growth factor-I gene: evidence for differential regulation in the development kidney. *Endocrinology*, **131**, 2793–9.

Shimatsu, A. & Rotwein, P. (1987*a*). Mosaic evolution of the insulin-like growth factors. Organization, sequence and expression of the rat insulin-like growth factor I gene. *Journal of Biological Chemistry*, **262**, 7894–900.

Shimatsu, A. & Rotwein, P.S. (1987*b*). Sequence of two rat insulin-like growth factor I precursers. *Nucleic Acids Research*, **15**, 7196.

Strauss, D.S. & Takemoto, C.D. (1991). Specific decrease in liver insulin-like growth factor-I and brain insulin-like growth factor-II gene expression in energy-restricted rats. *Journal of Nutrition*, **121**, 1279–86.

Sussenbach, J.S., Steenbergh, P.H. & Holthuizen, P. (1992). Structure and expression of the human insulin-like growth factor genes. *Growth Regulation*, **2**, 1–9.

Thissen, J.-P. & Underwood, L.E. (1992). Translational status of the insulin-like growth factor-I mRNAs in liver of protein-restricted rats. *Journal of Endocrinology*, **132**, 141–7.

Thomas, M.J., Kikuchi, K., Bichell, D.P. & Rotwein, P. (1994). Rapid activation of rat insulin-like growth factor-I gene transcription by growth hormone reveals no alterations in deoxyribonucleic acid-protein interactions within the major promoter. *Endocrinology*, **135**, 1584–92.

Thomas, M.J., Kikuchi, K., Bichell, D.P. & Rotwein, P. (1995). Characterization of deoxyribonucleic acid–protein interactions at a growth hormone-inducible nuclease hypersensitive site in the rat insulin-like growth factor-I gene. *Endocrinology*, **136**, 562–9.

Tollet, P., Enberg, B. & Mode, A. (1990). Growth hormone (GH) regulation of cytochrome P-450 11C12, insulin-like growth factor-I

(IGF-I) and GH receptor messenger RNA expression in primary rat hepatocytes: a hormonal interplay with insulin, IGF-I and thyroid hormone. *Molecular Endocrinology*, **4**, 1934–42.

Tomas, F.M., Knowles, S.E., Chandler, C.S., Francis, G.L., Owens, P.C. & Ballard, F.J. (1993*a*). Anabolic effects of insulin-like growth factor-I (IGF-I) and an IGF-I variant in normal female rats. *Journal of Endocrinology*, **137**, 413–21.

Tomas, F.M., Knowles, S.E., Owens, P.C., Chandler, C.S., Francis, G.L. & Ballard, F.J. (1993*b*). Insulin-like growth factor-I and more potent variants restore growth of diabetic rats without inducing all characteristic insulin effects. *Biochemical Journal*, **291**, 781–6.

VandeHaar, M.J., Moats-Staats, B.M., Davenport, M.L., Walker, J.L., Ketelslegers, J-M, Sharma, B.K. & Underwood, L.E. (1991). Reduced serum concentrations of insulin-like growth factor-I (IGF-I) in protein-restricted rats are accompanied by reduced IGF-I mRNA levels in liver and skeletal muscle. *Journal of Endocrinology*, **130**, 305–12.

Wong, E.A., Ohlsen, S.M., Godfredson, J.A., Dean, D.M. & Wheaton, J.E. (1989). Cloning of ovine insulin-like growth factor-I cDNAs: heterogeneity in the mRNA population. *DNA*, **8**, 649–57.

Zapf, J., Froesch, E.R. & Humbel, R.E. (1981). The insulin-like growth factors (IGFs) of human serum–chemical and biological characterization and aspects of their possible physiological role. *Current Topics in Cell Regulation*, **19**, 257–309.

Zarrilli, Z., Bruni, C.B. & Riccio, A. (1994). Multiple levels of control of insulin-like growth factor gene expression. *Molecular and Cellular Endocrinology*, **101**, R1–14.

DAVID J. HILL

Growth factor interactions in epiphyseal chondrogenesis

Introduction

Cartilage is the template on which much of the skeletal bone is formed, both during the primary ossification processes of embryonic and fetal development, and during pre- and post-natal longitudinal skeletal growth as a result of epiphyseal chondrogenesis. Three main classes of growth factors have been associated with these processes, the fibroblast growth factors (FGFs), the insulin-like growth factors -I and -II (IGFs-I and -II), and members of the transforming growth factor-β (TGF-β) family, including TGF-βs1, 2 and 3, and bone morphogenetic proteins 2B and 3. This paper reviews three likely roles of these growth factors in the regulation of limb and skeletal formation, and in the subsequent processes of epiphyseal chondrogenesis.

Limb development

Formation of the limb buds and the subsequent skeletal structure of the limbs have been well studied in the chick embyro, and are now thought to be tightly controlled by peptide growth factors. The limb buds first develop as a thickening of the body wall mesenchyme, the surface ectoderm of which is induced by the underlying mesenchyme to form a specialized structure called the apical ectodermal ridge. The mesenchyme beneath the apical ectodermal ridge is maintained in an undifferentiated, rapidly proliferating state and enables outgrowth of the limb to occur. Limb outgrowth is promptly arrested following removal of the apical ectodermal ridge. As mesenchyme moves distally to the progress zone, so it undergoes a condensation and morphogenic change to become cartilage. Sub-periosteal bone then develops on the surface of the cartilage immediately below the perichondrium to give rise to primary ossification structures. Increase in length of the long bones then continues by epiphyseal chondrogenesis and subsequent ossification. Superimposed upon this sequence of outgrowth and differentiation is the formation of the pentadactyl pattern of the limb in the dorso-ventral plane. This is controlled by a diffusible morphogen which is released within a specialized area of mesenchyme on the ventral aspect of the progress zone called the polarizing region, whose actions include the sequential activation of homeobox genes.

In the rat embryo, Beck *et al.* (1987) localized IGF-II mRNA to pre-cartilaginous mesenchymal condensations, perichondrium and immature chondrocytes, in addition to the periosteum and centres of intramembraneous ossification. Both IGF-I and -II mRNAs were also localized to limb bud mesenchyme in the rat fetus by Streck *et al.* (1992), who showed additionally that, while IGF-II expression was strongest in the presumptive skeleton and muscle at the centre of the limbs, IGF-I mRNA was absent from these areas but strongly expressed in the peripheral mesenchyme beneath the epithelium. Neither IGF isomer was strongly expressed in the rapidly dividing mesenchymal cells of the progress zone. While IGF-II is a potent mitogen for isolated limb bud mesenchyme from the rat *in vitro* (Bhaumick & Bala, 1991), the above findings suggest that the role of IGFs is not primarily as mitogens, but involves the initiation of differentiation pathways for skeletal and muscular elements. In the first trimester human fetus, also, IGF-II mRNA was abundant in perichondrial areas (Han, D'Ercole & Lund, 1987). We localized IGF peptides by immunocytochemistry in the chick embryo limb buds (Ralphs, Wylie & Hill, 1990). At stages 20–24 a uniform presence of IGFs was seen in undifferentiated mesenchyme, but this disappeared in the pre-chondrogenic areas of condensation. As chondrocytes appeared around stage 28, so immunoreactive staining for IGFs returned. By stage 36 endochondral calcification had begun and intense IGF staining was associated with hypertrophic chondrocytes, as well as osteoblasts in the sub-periosteum of the membraneous bone.

During the early development of the limb bud, IGFBP-2 expression is seen within the anterior-posterior strip of ectoderm which will become the apical ectodermal ridge, and IGFBP-2 continues to be expressed here until outgrowth is complete (Streck *et al.*, 1992). It is possible that the function of IGFBP-2 in the apical ectodermal ridge is to negate an IGF-II dependent drive towards differentiation in the underlying progress zone mesenchyme, and to maintain a stem cell population.

TGF-β1, 2 and 3 isomers are all expressed within the developing skeleton of the mouse embryo. Heine *et al.* (1987) observed the distribution of TGF-β1 peptide by immunocytochemistry from 11 to 18 days gestation. Strong staining was seen in all mesenchyme undergoing condensation and cartilage formation, which persisted during ossification within the newly formed osteoblasts. Analysis of TGF-β1 mRNA distribution by *in situ* hybridization revealed a high expression in perichondrial osteocytes involved with membraneous calcification (Lehnert & Akhurst, 1988). However, differentiated cartilage contained little TGF-β1 mRNA, although this is a site of peptide synthesis. A similar pattern has been described for TGF-β2 and -β3 in developing skeletal tissues (Pelton *et al.*, 1990). In the human embryo of 32–57 days gestation, TGF-β2 and -β3 mRNAs were localized to chondrogenic areas (Gatherer *et al.*, 1990). TGF-β2 mRNA was located within the pre-cartilaginous blastoma of limb bud mesenchyme, and in later development in actively proliferating chondroblasts at the epiphyseal/diaphyseal boundary. Messenger RNA for TGF-β3 was first seen in the developing intervertebral discs and in the perichondrium of cartilage associated with the vertebral column, but not with long bones. An intense site of TGF-β1 abundance was in areas of membraneous bone formation and, in fetuses of 10–12 weeks gestation, in osteogenic cells at sites of endochondrial calcification in the long bones. No TGF-β mRNA was located in the hypertrophic chondrocytes which immediately precede the area of provisional calcification. A divergent expression of TGF-β isoforms in skeletal primordia suggests distinct biological roles, and that of TGF-β2 would support a role in cartilaginous induction. Evidence for this is provided by the observations that mammalian TGF-β1 and -β2 induced the appearance of phenotypic chondrocytes, associated with increased sulphated mucopolysaccharide and type II collagen synthesis, in chick embryo mesenchyme cultures. Other members of the TGF-β family, namely bone morphogenetic proteins-2B and -3 (osteogenin), will also induce cartilage formation from embryonic mesoderm in the chick (Carrington *et al.*, 1991; Chen *et al.*, 1991) making the identity of the endogenous active ligand unclear. When TGF-β1 or -β2 was injected into the sub-periosteal region of the femurs from newborn rats, local intramembraneous bone and cartilage formation resulted (Joyce *et al.*, 1990). After injections were terminated, the new cartilage underwent endochondrial calcification. These results strongly suggest that TGF-β isomers are key players in the formation of cartilage from undifferentiated mesenchyme, and in the subsequent primary ossification process.

FGF2 was present at both mRNA and protein levels during limb bud formation in the chick and mouse (Munaim, Klagsbrun & Toole, 1988; Herbert *et al.*, 1990), peptide levels in the chick limb being greatest on day 3 of gestation (stage 18) when the cell proliferation rate was highest. During this rapid proliferation mesenchymal cells secrete an extracellular matrix rich in hyaluronic acid. Using isolated chick limb bud cells, FGF2 was shown to potentiate hyaluronic acid release to form pericellular coats (Munaim, Klagsbrun & Toole, 1991). A loss of hyaluronic acid synthesis *in vitro* coincided with the timing of condensation of mesoderm into the chondrogenic and myogenic regions of the limb bud, and a decline in FGF2 abundance, at stages 22–26. Within the mouse embryo, Gonzalez *et al.* (1990) used immunocytochemistry to localize FGF2 at 18 days gestation. Positive staining was apparent in chondrocytes of the hyaline cartilage and in the perichondrium. Within ossification centres FGF2 was absent from hypertrophic cells but present within the extracellular matrix, osteoblasts and vascular endothelial cells. A different experimental approach was that of Liu and Nicholl (1988) who transplanted fetal rat paws, harvested on day 10 of gestation, under the kidney capsule of adult hosts which were then infused with FGF2 or anti-FGF2 antiserum via the renal artery. Infusion of FGF2 antiserum significantly retarded the growth of the explants and their ossification. Conversely, paw size was increased by administration of FGFs. Other isomers of FGF may also be involved in limb development, FGF5 mRNA appearing within the limb mesenchyme between embryonic days 12.5 and 14.5 in mouse (Haub & Goldfarb, 1991). Expression was limited to a patch of cells ventral to the presumptive femurs which was undergoing cartilage formation. The above evidence suggests a mitogenic role for FGF2 in mesenchyme proliferation, and a possible morphogenetic role for FGF5 during cartilage induction. However, the strongest evidence linking FGFs to limb formation does not involve the mesenchyme but the apical ectodermal ridge.

As soon as the apical ectodermal ridge is formed, at day 10 of gestation in the mouse, a high expression of FGF4 mRNA is seen in the posterior half, and expression persists until day 12 (Niswander & Martin, 1992). Several members of the FGF family can substitute for the apical ectodermal ridge and maintain limb bud outgrowth *in vitro* in both the mouse and chick (Niswander & Martin, 1992, 1993), suggesting that an FGF is involved in the endogenous, signalling between the epithelium and the underlying mesenchyme. Recently, Niswander *et al.* (1993) demonstrated that recombinant FGF4 could substitute for the ridge *in ovo*, and not only maintain limb bud outgrowth but signal the correct spatial information to achieve normal pattern formation. This would imply that FGF4 is capable of regulating an appropriate release of morphogens from the polarizing zone within the mesenchyme. Contradictory evidence was provided by Fallon *et al.* (1994) who showed that only FGF2 was detectable in the chick limb bud, and that exogenous FGF2 could substitute for the apical ectomermal ridge.

It is now possible to predict which receptor types are involved in FGF signalling within both mesenchyme and ectoderm in the developing limbs. In the mouse embryo, the FGFR2 receptor was first expressed on day 9.5 in limb bud mesenchyme, with a concentration gradient increasing in a posterior and proximal direction. At this time the expression of FGFR1 was more diffuse than that of FGFR2, within the limb bud mesenchyme, the somites and organ rudiments (Peters *et al.*, 1992). By day 11.5 FGFR2 mRNA was localized to mesenchymal aggregates corresponding to the future bones, and in the surface ectoderm of the limb, being strongest in the interdigital web. At day 12.5 gestation FGFR2 mRNA located to chondrification centres, and at 14.5 to the bodies of the distal bones. This temporal pattern of expression strongly suggests that FGFR2 mediates FGF actions on the chondrogenic pathways, while FGFR1 may mediate FGF actions on the surrounding undifferentiated mesenchyme. The FGFR4 receptor mRNA was found by *in situ* hybridization to map to areas of cartilage condensation, while FGFR3 mRNA was abundant in the resting cartilage during the subsequent process of endochondral calcification (Stark, McMahon & McMahon, 1991; Peters *et al.*, 1993).

Epiphyseal chondrogenesis

Longitudinal skeletal growth arises from new bone formation as a result of epiphyseal chondrogenesis, and subsequent replacement of cartilage by bone. Adjacent to the epiphysis is a stem cell population of precursor chondrocytes, which gives rise to a zone of closely packed and highly proliferative chondrocytes. After a number of cell replications, cells are pushed towards the diaphysis within longitudinally arranged columns. Proliferative activity gradually decreases while the cells hypertrophy and increase their rate of macromolecular synthesis. As matrix synthesis increases, the cells become postmitotic and terminally differentiated. Mineralization then occurs between the columns of chondrocytes, and the chondrocytes are replaced by macrophages and osteocytes.

Cells within the proliferative zone of the rat growth plate continuously cycle with a cell cycle length of approximately 54 h (Walker & Kember, 1972). They also secrete an extracellular matrix consisting mainly of type-II collagen fibrils and sulphated glycosaminoglycans, the most abundant of which are chondroitin-4 and -6 sulphates, and hyaluronic acid. Cells within the hypertrophic zone have a decreased nuclear labelling index, but an increased rate of collagen and sulphated mucopolysaccharide synthesis, with an increased ratio of protein to DNA. Increases or decreases in growth rate are predominantly due to alterations in hypertrophic activity rather than changes in chondrocyte proliferation rate (Hunziker, 1988). Terminal differentiation in the lower hypertrophic zone is characterized by a dramatic fall in mucopolysaccharide synthesis but a maintenance of collagen synthesis, the expression of a distinctive type-X collagen, and the appearance of alkaline phosphatase activity. Adjacent to the diaphysis, cartilage matrix is eroded and replaced by hydroxyapatite at a similar rate to new cartilage formation as a result of proliferation and hypertrophy. The length of the diaphysis, therefore, increases while the width of the growth plate remains constant. Calcification is initiated in matrix vesicles in the lower hypertrophic zone coincident with vascular invasion by capillary sprouting, and the delivery of monocytes and chondroclasts which degrade cartilage matrix and may destroy hypertrophic chondrocytes.

Growth factors may therefore influence epiphyseal chondrogenesis by altering the rate of chondrocyte proliferation, by modulation of the synthetic rate of extracellular matrix molecules, or by changing the rate of terminal differentiation and calcification. There is substantial evidence that FGFs, IGFs and TGF-β isomers are expressed locally within the growth plates during both prenatal and postnatal epiphyseal chondrogenesis.

The presence of FGF2 within cartilage has been recognized for over 10 years (Bekoff & Klagsbrun, 1982), and its production by isolated chick growth plate chondrocytes has been described (Rosier, Landesberg & Puzas, 1991). Extensive analysis of IGF expression, control, and action within epiphyseal growth plate cartilage has provided conclusive proof for a substantial paracrine component of control in postnatal longitudinal skeletal growth. This series of studies was initiated by Isaksson and co-workers who found that pituitary growth hormone (GH) promoted the replication of isolated chondrocytes *in vitro*, or caused a widening of the epiphyseal growth plate of the proximal tibia when administered locally (Isaksson, Jansson & Gause, 1982; Lindahl *et al.*, 1986). The sophistication of the model was enhanced by infusion of GH into a single femoral artery using an osmotic minipump. This, again, resulted in a local growth response in the treated limb compared to the contralateral control limb (Nilsson *et al.*, 1987).

With the demonstration that IGF I peptide could be visualized, using immunocytochemistry, in association with proliferative and maturing chondrocytes of the growth plates in rat long bones (Andersson *et al.*,

1986) Isaksson performed a definitive experiment. When longitudinal sections were examined from the growth plates of hypophysectomized rats chondrocytes no longer demonstrated immunolocalization of IGF I, but abundant staining returned when hypophysectomized rats were treated with GH (Nilsson *et al.*, 1986). These experiments were repeated by others and extended. The ability of locally infused GH to produce a localized epiphyseal growth response in rat could be reversed by the simultaneous infusion of blocking antibody against IGF I (Schlechter *et al.*, 1986). The model that emerged has shown IGF I to be associated with the proliferative chondrocytes of the growth plate, and that its presence mediates the growth promoting actions of GH *in vivo*. Direct proof that GH controlled the transcription of the IGF I gene in post-natal epiphyseal growth plate chondrocytes quickly followed (Isgaard *et al.*, 1988). In agreement with the hypothesized growth system the distribution of GH receptors, determined by immunocytochemistry, was found to be limited to the stem cell and proliferative chondrocytic population of the rabbit growth plate (Barnard *et al.*, 1988), while IGF receptors were most abundant in the proliferative region but were also found on hypertrophic chondrocytes (Trippel, Van Wyk & Mankin, 1986). The latter observation is in agreement with the observed biological actions of IGFs *in vitro*, which include a predominantly mitogenic action in the proliferative chondrocyte region, and a stimulation of mucopolysaccharide synthesis in the post-mitotic hypertrophic chondrocytes adjacent to newly formed bone (Hill, 1979).

The anabolic actions of GH on the epiphyseal growth plate may not be mediated entirely by locally produced IGF I, since GH may synergize with IGF I in addition to controlling its expression. Studies with the clonal growth of isolated chondrocytes in soft agar showed that the morphology of colonies induced by the addition of GH differed markedly from those resulting from exposure to IGF I (Lindahl, Nilsson & Isaksson, 1987). Similarly, while IGF I stimulated both DNA synthesis and proteoglycan synthesis by bovine articular chondrocytes a much enhanced response was seen following the addition of GH. The latter peptide had no effects in the absence of IGF I (Smith *et al.*, 1989).

Using *in situ* hybridization, TGF-β1 mRNA was shown to be expressed in abundance in bone cells of the human fetus in the first trimester, although relatively little was present within the growth plate (Sandberg *et al.*, 1988). However, TGFβ immunoreactivity was seen in chondrocytes and cartilage matrix of fetal and post-natal mice (Ellingsworth *et al.*, 1986; Heine *et al.*, 1987), while bio-detectable TGF-β was released from isolated growth plate chondrocytes from the post-natal chick, this being greater for hypertrophic cells than for cells of the proliferative zone (Rosier *et al.*, 1989; Gelb, Rosier & Puzas, 1990). TGF-β may be stored within the matrix of bovine articular cartilage in a relatively inactive form, perhaps as the latent precursor molecule, which could then be activated by locally produced proteases.

We have studied the regulation of epiphyseal chondrogenesis in the fetus as a model of peptide growth factor interaction during cell proliferation and differentiation, our model being the proximal tibia of the ovine fetus. FGF2 and its high affinity receptor, FGFR1, were strongly expressed at both mRNA and peptide levels in the proliferative chondrocyte zone of the ovine growth plate, decreased during cell differentiation, and were absent from the hypertrophic chondrocytes. No IGF-I mRNA was observed in the fetal growth plate, but IGF-II mRNA and peptide was predominantly associated with the differentiating, but still mitotically active, chondrocytes. Unlike IGF-I in postnatal life, the expression and release of IGF-II in fetal growth plate chondrocytes was not increased in response to growth hormone. An associated expression of IGFBP-2 and IGFBP-3 was found. As chondrocytes began to hypertrophy, so mRNA for IGF-II declined and that encoding TGF-β1 appeared. The terminally differentiated chondrocytes also expressed IGFBP-5, which was not seen in other areas of the growth plate. In summary, as chondrocytes passed from proliferation to differentiation to hypertrophy, they sequentially expressed FGF2, then IGF-II, and finally TGF-β1. This anatomical distribution allowed testable hypotheses to be formulated with regard to growth factor contribution to epiphyseal chondrogenesis.

The expression of FGF2 and its receptor in stem and proliferating chondrocyte populations suggests a role as an autocrine mitogen. Using isolated chondrocyte cultures FGF2 was found to be released and to contribute to DNA synthesis, while a neutralizing antibody against FGF2 decreased endogenous DNA synthesis in cells by 50% (Hill & Logan, 1992*a*; Hill *et al.*, 1992*a*). Exogenous FGF-2 was 100–500 times more potent than IGF-I, IGF-II or insulin as a mitogen for fetal growth plate chondrocytes (Hill & Logan, 1992*b*; Hill *et al.*, 1992*b*). Insulin had an equivalent mitogenic action to IGF-I at low nanomolar concentrations (Hill & De Sousa, 1990), and both were an order of magnitude more effective, on a concentration basis, than was IGF-II, the endogenously produced IGF isomer. Conversely, IGF-II, which was expressed by maturing chondrocytes, was a potent stimulator of glycosaminoglycan and collagen synthesis by chondrocytes, parameters of a differentiated phenotype (Hill *et al.*, 1992*a*). It is likely that the actions of endogenous IGF-II on chondrocyte growth and maturation are modulated by endogenous IGFBPs. We found that exogenous IGFBP-2 had a biphasic effect on IGF-II stimulated chondrocyte DNA synthesis, enhancing the actions of IGF-II at concentrations which were approximately equimolar to the added growth factor, but inhibiting IGF-II action when the IGFBP was present in excess. At lower concentrations the IGFBP-2 may bind to cell surface integrins via its concensus RGD binding sequence, and may concentrate IGF-II at the cell surface where it is readily accessible to the high affinity type-1 receptors. At greater concentrations of IGFBP-2 the integrin binding sites may be fully occupied, and soluble IGFBP-2 may then directly compete with the signalling receptor for ligand binding.

Finally, TGF-β1, which was located in terminally differentiated cells, inhibited chondrocyte replication in response to other mitogens yet potentiated extracellular matrix molecule production. TGF-β1 increased sulphated mucopolysaccharide synthesis and enhanced collagenous protein synthesis at the expense of non-collagenous protein (Hill *et al.*, 1992*a*). Many of the biochemical features of epiphyseal chondrogenesis might therefore be explained by interactions between endogenously produced peptide growth factors. It seems likely that mineralization involves the further interaction with thyroxine, which was found to reverse the mitogenic actions of IGFs on chondrocytes while inducing alkaline phosphatase production, a marker of terminal differentiation (Ohlsson *et al.*, 1992).

Acknowledgements

Those studies performed in the author's laboratory were funded by the Medical Research Council of Canada.

References

Andersson, I., Billig, H., Fryklund, L., Hansson, H-H., Isaksson, O., Isgaard, J., Nilsson, A., Rozell, B., Skottner, A. & Stemme, S. (1986). Localization of IGF I in adult rats. Immunohistochemical Studies. *Acta Physiologica Scandinavica*, **126**, 311–12.

Barnard, R., Haynes, K.M., Werther, G.A. & Waters, M.J.. (1988). The ontogeny of growth hormone receptors in the rabbit tibia. *Endocrinology*, **122**, 2862–9.

Beck, F., Samani, N.J., Penschow, J.D., Thorley, B., Tregear, G.W. & Coghlan, J.P. (1987). Histochemical localization of IGF-I and -II mRNA in the developing rat embryo. *Development*, **101**, 175–84.

Bekoff, M.C. & Klagsbrun, M. (1982). Characterization of growth factors in human cartilage. *Journal of Cell Biochemistry*, **20**, 233–45.

Bhaumick, B. & Bala, R.M. (1991). Differential effects of insulin-like growth factors I and II on growth, differentiation and glucoregulation in differentiating chondrocyte cells in culture. *Acta Endocrinologica*, **125**, 201–11.

Carrington, J.L., Chen, P., Yanagishita, M. & Reddi, A.H. (1991). Osteogenin (Bone morphometric protein-3) stimulates cartilage formation by chick limb bud cells *in vitro*. *Developmental Biology*, **146**, 406–15.

Chen, P., Carrington, J.L., Hammonds, R.G. & Reddi, A.H. (1991). Stimulation of chondrogenesis in limb bud mesoderm cells by recombinant human bone morphometric protein 2B (BMP-2B) and modulation by transforming growth factor β_1 and B_2. *Experimental Cell Research*, **195**, 509–15.

Elingsworth, L.R., Brennan, J.E., Fok, K., Rosen, D.M., Bentz, H., Piez, K.A. & Seyedin, S.M. (1986). Antibodies to the *N*-terminal portion of cartilage-inducing factor A and transforming growth factor β. *Journal of Biological Chemistry*, **261**, 12362–7.

Fallon, J.F., Lopez, A., Ros, M.A., Savage, M.P., Olwin, B.B. & Simandl, B.K. (1994). FGF-2: Apical ectodermal ridge growth signal for chick limb development. *Science*, **264**, 104–6.

Gatherer, D., TenDijke, P., Baird, D.T. & Akhurst, R.J. (1990). Expression of TGF-*β* isoforms during first trimester human embryogenesis. *Development*, **110**, 445–60.

Gelb, D.E., Rosier, R.N. & Puzas, J.E. (1990). The production of transforming growth factor-*β* by chick growth plate chondrocytes in short term monolayer culture. *Endocrinology*, **127**, 1941–7.

Gonzalez, A-M., Buscaglia, M., Ong, M. & Baird, A. (1990). Distribution of basic fibroblast growth factor in the 18-day rat fetus: localization in the basement membranes of diverse tissues. *Journal of Cell Biology*, **110**, 753–65.

Han, V.K.M., D'Ercole, A.J. & Lund, P.K. (1987). Cellular localization of somatomedin (insulin-like growth factor) messenger RNA in the human fetus. *Science*, **236**, 193–7.

Haub, O. & Goldfarb, M. (1991). Expression of fibroblast growth factor-5 gene in the mouse embryo. *Development*, **112**, 397–406.

Heine, U.I., Munoz, E.F., Flanders, K.C., Ellingsworth, L.R., Lam, H.Y.P., Thompson, N.L., Roberts, A.B. & Sporn, M.B. (1987). Role of transforming growth factor-*β* in the development of the mouse embryo. *Journal of Cell Biology*, **105**, 2861–76.

Herbert, J.M., Basilico, C., Goldfarb, M., Haub, O. & Martin, G.R. (1990). Isolation of cDNAs encoding four mouse FGF family members and characterization of their expression patterns during embryogenesis. *Developmental Biology*, **138**, 454–63.

Hill, D.J. (1979). Stimulation of cartilage zones of the calf costochondral growth plate *in vitro* by growth hormone dependent rat plasma somatomedin activity. *Journal of Endocrinology*, **83**, 219–27.

Hill, D.J. & De Sousa, D. (1990). Insulin is a mitogen for isolated epiphyseal growth plate chondrocytes from the fetal lamb. *Endocrinology*, **126**, 2661–70.

Hill, D.J. & Logan, A. (1992*a*). Interactions of peptide growth factors during DNA synthesis in isolated ovine fetal growth plate chondrocytes. *Growth Regulation*, **2**, 122–32.

Hill, D.J. & Logan, A. (1992*b*). Cell cycle-dependent localization of immunoreactive basic fibroblast growth factor to cytoplasm and nucleus of isolated ovine fetal growth plate chondrocytes. *Growth Factors*, **7**, 215–31.

Hill, D.J., Logan, A., McGarry, M. & DeSousa, D. (1992*a*). Control of protein and matrix molecule synthesis in isolated ovine growth plate chondrocytes by the interactions of basic fibroblast growth factor, insulin-like growth factor I, insulin and transforming growth factor β. *Journal of Endocrinology*, **133**, 363–73.

Hill, D.J., Logan, A., Ong, M., DeSousa, D. & Gonzalez, A.M. (1992*b*). Basic fibroblast growth factor is synthesized and released by isolated ovine fetal growth plate chondrocytes: Potential role as an autocrine mitogen. *Growth Factors*, **6**, 277–94.

Hunziker, E.B. (1988). Growth plate structure and function. *Pathology and Immunopathology Research*, **7**, 9–13.

Isaksson, O., Jansson, J-O. & Gause, I.A. (1982). Growth hormone stimulates longitudinal bone growth directly. *Science*, **216**, 1237–9.

Isgaard, J., Moller, C., Isaksson, O., Nilsson, A., Mathews, L.S. & Norstedt, G. (1988). Regulation of insulin-like growth factor messenger ribonucleic acid in rat growth plate by growth hormone. *Endocrinology*, **122**, 1515–20.

Joyce, M.E., Roberts, A.B., Sporn, M.B. & Bolander, M.E. (1990). Transforming growth factor-β and the initiation of chondrogenesis and osteogenesis in the rat femur. *Journal of Cell Biology*, **110**, 2195–207.

Lehnert, S.A. & Akhurst, R.J. (1988). Embryonic expression pattern of TGF beta type-1 RNA suggests both paracrine and autocrine mechanisms of action. *Development*, **104**, 263–73.

Lindahl, A., Isgaard, J., Nilsson, A. & Isaksson, O. (1986). Growth hormone potentiates colony formation of epiphyseal chondrocytes in suspension culture. *Endocrinology*, **118**, 1843–8.

Lindahl, A., Nilsson, A. & Isaksson, O. (1987). Effects of growth hormone and insulin-like growth factor I (IGF-I) on colony formation of rabbit epiphyseal chondrocytes of different maturation. *Journal of Endocrinology*, **115**, 263–71.

Liu, L. & Nicoll, C.S. (1988). Evidence for a role of basic fibroblast growth factor in rat embryonic growth and differentiation. *Endocrinology*, **123**, 2027–31.

Munaim, S.I., Klagsbrun, M. & Toole, B.P. (1988). Developmental changes in fibroblast growth factor in the chicken embryo limb bud. *Proceedings of the National Academy of Sciences, USA*, **85**, 8091–3.

Munaim, S.I., Klagsbrun, M. & Toole, B.P. (1991). Hyaluronan-dependent pericellular coats of chick embryo limb mesoderm cells: induction by basic fibroblast growth factor. *Developmental Biology*, **143**, 297–302.

Nilsson, A., Isgaard, J., Lindahl, A., Dahlstrom, A., Skottner, A. & Isaksson, O. (1986). Regulation by growth hormone of number of chondrocytes containing IGF I in the rat growth plate. *Science*, **233**, 571–4.

Nilsson, A., Isgaard, J., Lindahl, A., Peterson, L. & Isaksson, O. (1987). Effects of unilateral arterial infusion of GH and IGF I on tibial longitudinal bone growth in hypophysectomized rats. *Calcified Tissue International*, **40**, 91–6.

Niswander, L. & Martin, G.R. (1992). *Fgf-4* expression during gastrulation, myogenesis, limb and tooth development in the mouse. *Development*, **114**, 755–68.

Niswander, L. & Martin, G.R. (1993). FGF-4 and BMP-2 have opposite effects on limb growth. *Nature*, **361**, 68–71.

Niswander, L., Tickle, C., Vogel, A., Booth, I. & Martin, G.R. (1993). FGF-4 replaces the apical ectodermal ridge and directs outgrowth and patterning of the limb. *Cell*, **75**, 579–87.

Ohlsson, C., Nilsson, A., Isaksson, O., Bentham, J. & Lindahl, A. (1992). Effects of tri-iodothyronine and insulin-like growth factor-I (IGF-I) on alkaline phosphatase activity, [^{3}H]thymidine incorporation and IGF-I receptor mRNA in cultured rat epiphyseal chondrocytes. *Journal of Endocrinology*, **135**, 115–21.

Pelton, R.W., Dickinson, M.E., Moses, H.L. & Hogan, B.L.M. (1990). *In situ* hybridization analysis of TGFβ_3 RNA expression during mouse development: comparative studies with TGFβ_1 and β_2. *Development*, **110**, 609–20.

Peters, K., Ornitz, D., Werner, S. & Williams, L. (1993). Unique expression pattern of the FGF receptor 3 gene during mouse organogenesis. *Developmental Biology*, **155**, 423–30.

Peters, K.G., Werner, S., Chen, G. & Williams, L.T. (1992). Two FGF receptor genes are differentially expressed in epithelial and mesenchymal tissues during limb formation and organogenesis in the mouse. *Development*, **114**, 233–43.

Ralphs, J., Wylie, L. & Hill, D.J. (1990). Distribution of insulin-like growth factor peptides in the developing chick embryo. *Development*, **109**, 51–8.

Rosier, R.N., Landesberg, R.L. & Puzas, J.E. (1991). An autocrine feedback loop for regulation of TGF-β and FGF production in growth plate chondrocytes. *Journal of Bone and Mineral Research*, **6**, (suppl. 1) S211.

Rosier, R.N., O'Keefe, R.J., Crabb, I.D. & Puzas, J.E. (1989). Transforming growth factor-beta: an autocrine regulator of chondrocytes. *Connective Tissue Research*, **20**, 295–301.

Sandberg, M., Vuorio, T., Hirvonen, H., Alitalo, K.L. & Vuorio, E. (1988). Enhanced expression of TGF-β and *c-fos* mRNAs in the growth plates of developing human long bones. *Development*, **102**, 461–70.

Schlechter, N.L., Russell, S.M., Spencer, E.M. & Nicoll, C.S. (1986). Evidence suggesting that the direct growth-promoting effect of growth hormone on cartilage *in vivo* is mediated by local production of somatomedin. *Proceedings of the National Academy of Sciences, USA*, **83**, 7932–4.

Smith, R.L., Palathumpat, M.V., Ku, C.W. & Hintz, R.L. (1989). Growth hormone stimulates insulin-like growth factor I actions on adult articular chondrocytes. *Journal of Orthopaedic Research*, **7**, 198–207.

Stark, K.L., McMahon, J.A. & McMahon, A.P. (1991). FGFR-4, a new member of the fibroblast growth factor receptor family, expressed in the definitive endoderm and skeletal muscle lineages of the mouse. *Development*, **113**, 641–51.

Streck, R.D., Wood, T.L., Hsu, M-S. & Pintar, J.E. (1992). Insulin-like growth factor I and II and insulin-like growth factor binding protein-2 RNAs are expressed in adjacent tissues within rat embryonic and fetal limbs. *Developmental Biology*, **151**, 586–96.

Trippel, S.B., Van Wyk, J.J. & Mankin, H.J. (1986). Localization of somatomedin-C binding to bovine growth-plate chondrocytes *in situ*. *Journal of Bone and Joint Surgery*, **68-A**, 897–903.

Walker, K.V.R. & Kember, N.F. (1972). Cell kinetics of growth cartilage in the rat tibia. I. Measurements in young male rats. *Cell and Tissue Kinetics*, **5**, 401–8.

ANN LOGAN

Developmental changes in the CNS response to injury: growth factor and matrix interactions

Introduction

After a penetrating injury, neurons of the adult mammalian central nervous system (CNS) show only a limited and transient ability to regenerate, so that any motor and sensory deficits incurred are permanent. This is in direct contrast to the situation in the fetal and perinatal CNS, where neurons show a remarkable growth capacity after injury, with negligible consequent functional deficits. Our laboratory has been studying the differences in cellular and trophic responses that underlie the developmental changes in the CNS wounding response.

The cellular response to injury

The mature CNS

The response to injury in the adult CNS is characterized by three sequential and overlapping events: acute haemorrhage and inflammation, followed by glial/collagen scar formation, which is accompanied by an abortive regeneration response by axotomized neurons (Maxwell *et al.*, 1990*a*). Briefly, at first the lesion is haemorrhagic and extravasated platelets play a major role in initiating both clotting in the lesion lumen and vasoconstriction in the wound edges associated with oedema and necrosis. Polymorphs, monocytes and later macrophages appear in large numbers. Their concentration in the wound is probably controlled by a homing response organized both by platelet factors and also by the expression of addressins on the endothelium of the brain vasculature and counter-receptors on leukocytes, as elsewhere in the body, although little is known of the details of this process in the damaged CNS. By 3–5 days, most of the extravasated erythrocytes have disappeared. The wound and its margins are filled with macrophages, monocytes and a few polymorphs. Fibroblasts and collagen fibres first appear at this stage to form a mesenchymal core in which extracellular matrix molecules, such as laminin and fibronectin, are deposited. The necrotic neuropil at the mesenchyme/CNS parenchymal interface is also absorbed over the 3–5 day period and large numbers of reactive astrocytes, with upregulated glial fibrillary acidic protein (GFAP) expression, appear in the CNS surrounding the wound. Microglia, which become reactive at about 2 days, also accumulate in the same peri-lesion neuropil. Scar tissue is laid down between 5 and 8 days. The cytoplasmic processes of astrocytes become concentrated at the CNS/mesenchymal boundary, form a continuous multilayered sheet and become locked together by tight junctions. Between these astrocytic processes and the mesenchymal core, a basal lamina is deposited, probably by the astrocytes under the influence of fibroblasts. Macrophages are lost from the mesenchymal core, but much fibrous collagen and fibronectin is deposited by fibroblasts.

Scarring essentially establishes a glia limitans in the wound (identical to and continuous with that of the external limiting membrane), composed of the mesenchymal elements of the leptomeninges externally, a basal lamina and astrocytic end-feet bound together by tight junctions internally. Fibroblasts in the central core of the lesion migrate in from the meninges. Thus, the scar matures sequentially from pial surface to the depths of the lesion. By 8–14 days, the scar is fully formed, the central mesenchymal fibrous core contracted and the pallasades of astrocyte processes compacted at the lesion margins. GFAP-activity declines, except at the wound margins (Fig. 1). A small number of macrophages remain in the core, but none is seen in the surrounding CNS tissue.

Embryonic/neonatal CNS

The mature response to injury described above is acquired neonatally, in the rat for example, between 8 and 12 days *post partum* (dpp) (Berry *et al.*, 1983; Maxwell, *et al.*, 1990*b*). Before 8 dpp, no mesenchymal elements accumulate in the wound and although some astrocytes become reactive, a glia limitans is not formed and the neuropil grows together obliterating all signs of the original lesion (Fig. 2). Typical scarring first appears at 8 dpp subpially and, over the next four days, invades the depths of the wound. Although the microglial response is qualitatively mature in the neonate, macrophage and fibroblast invasion from the pia into the wound is minimal. Curiously, however, the glial limitans externa is repaired after injury before 8 dpp. Clearly, understanding the biology of scar acquisition in the perinatal CNS

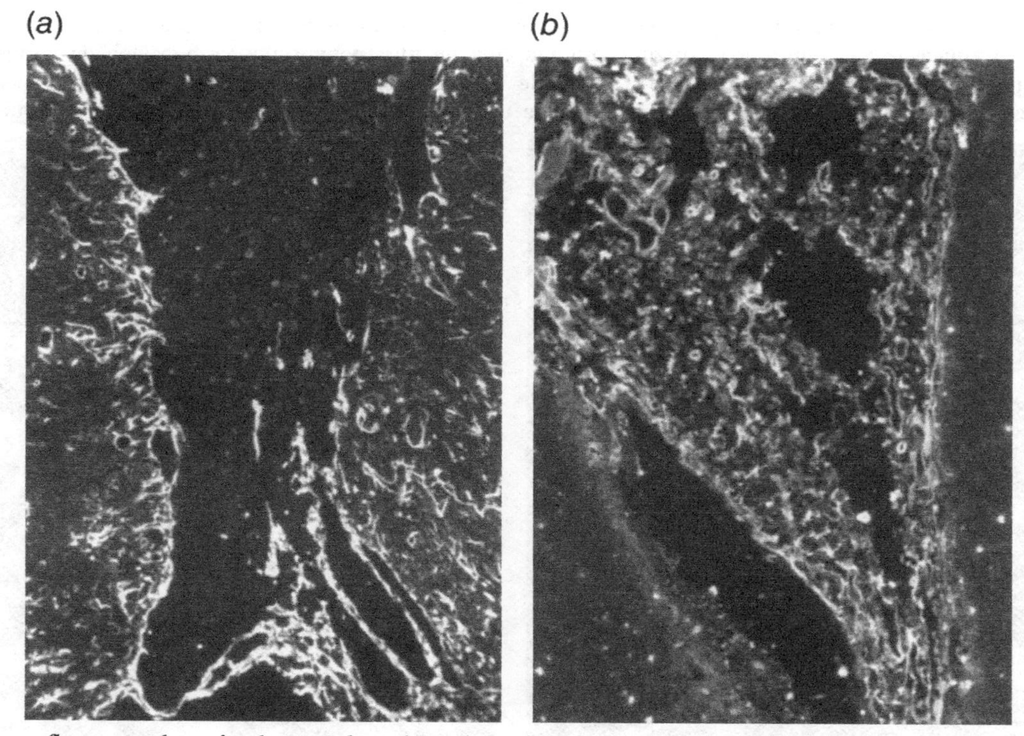

Fig. 1. Immunofluorescently stained coronal sections through lesion sites in the adult cerebral cortex, 10 days after injury. (a) shows GFAP-positive reactive astrocytes. These are organizing along the wound margins to form a limiting glial membrane (glia limitans). (b) shows an adjacent section stained with an anti-fibronectin antibody to demonstrate the fibrous matrix deposited in the core of the wound.

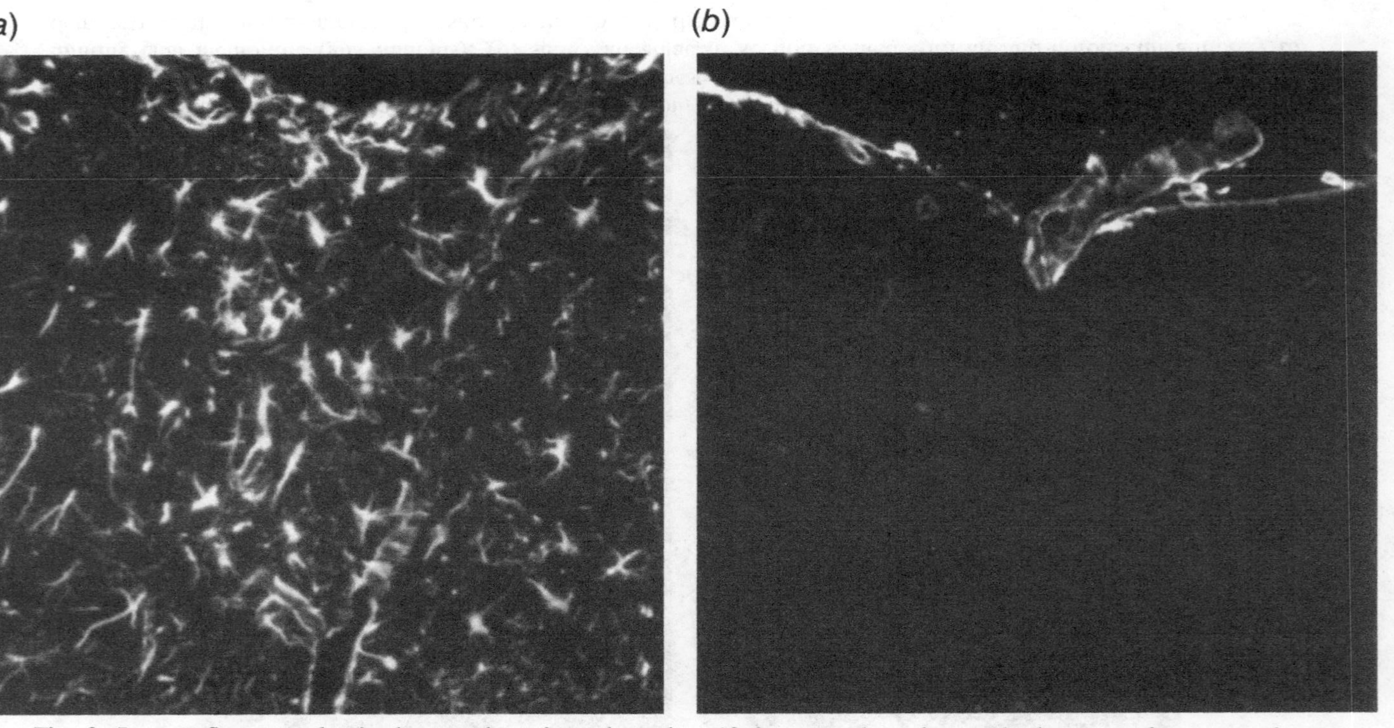

Fig. 2. Immunofluorescently sized coronal sections through a 10-day wound to the cerebral cortex of a neonatal rat. (a) shows GFAP-positive reactive astrocytes with no apparent organization into a glia limitans within the wound. (b) shows an adjacent section stained with an anti-laminin antibody, confirming the resealing of the pial glia limitans, but no basement membrane material within the wound and no fibrous scar.

could lead to a pharmacological strategy aimed at replicating the neonatal response to injury in the adult.

Degeneration/regeneration of neurons and their processes

Nerves in the injured embryonic and perinatal CNS show a very good capacity for growth and lesion sites are rapidly traversed by axons which make appropriate and functional distal synapses. However, it is not known whether the neuronal response represents true regeneration or *de novo* growth of neuroblasts. This is in direct contrast to the classical response observed for mature neurons to injury. In the adult there is an initial sprouting reaction from the cut axonal terminals, which is followed by degeneration. By 16 days after injury, all newly grown processes have died back to their original parent axons, a large number of which remain *in situ* in perpetuity, making no further attempt to grow (Cajal, 1928). It is not known if this abortive growth response of axons is also exhibited by dendrites. It is not known also why axons fail to sustain growth after injury. An early explanation was that growth arrest is scar related, since the cicatrix constitutes a physical barrier through which axons are unable to pass. But neuromata never form around a scar and, accordingly, this idea has received little support. None the less, it is clear that if regeneration becomes a possibility, concomitant inhibition of scarring is essential if axons are to restore disconnected functional pathways.

A contemporary explanation for the failure of regeneration in the CNS is that glia, probably both astrocytes and oligodendrocytes (Berry *et al.*, 1983), have surface membrane ligands which, when engaged by specific axonal membrane receptors, trigger the inhibition of growth by a Ca^{2+}-dependent intracellular signal. This hypothesis predicts successful regeneration in sites where either axonal receptors, glia ligands or both are not expressed. For example, the florid regeneration of the axons from fetal brain grafts implanted into adult brain (Wictorin *et al.*, 1992) is probably explained if embryonic neurons do not express the receptor for the ligand on mature host glia about the transplant site in the adult brain.

It is generally held that neurons have an absolute dependency on trophic molecules for growth and survival. It is possible that, in the adult CNS, viability of neurons is maintained by a continuous flow of neurotrophic molecules derived from the target and retrogradely transported along the axon to the perikaryon. Some neuronal cell death always follows axonal damage. In neonates, the parent cell bodies are particularly sensitive to axotomy. In embryonic and neonatal retina, for example, almost all retinal ganglion cells die after transection of the optic nerve, whereas in adults 20–30% survive. Separating perikaryon from target might cause cell death if the neuron is thereby deprived of essential target derived trophic factors. Differential viability of neural centres in the adult may be explained by differences in the degree of multiple target innervation by collaterals.

Neural degeneration is probably augmented by the release of proteases from microglia amassing in the wound and surrounding neuropil in the acute post injury period (Thanos, 1991). Cell death may also be enhanced by activation of proteases by Ca^{2+} flooding into the wound area from the blood as the blood brain barrier is breached. The *N*-methyl-D-aspartate (NMDA) receptor-mediated cytotoxicity associated with excitatory amino acid release immediately after trauma also contributes to neuronal cell death.

Definition of trophic mechanisms of the injury response

The spatio-temporal trophic cascade that results from invasive injuries of the CNS, and which regulates the subsequent cellular responses, involves multiple paracrine regulatory factors. Using molecular hybridization and immunochemical techniques to analyse histological sections and extracts of damaged neural tissue, the individual elements of this cascade are beginning to be defined. Attempts to identify key neurotrophic and fibrogenic factors have led to the implication of two growth factors that have been well studied by us, basic fibroblast growth factor (FGF-2) and transforming growth factor-β1 (TGF-β1).

FGF-2 as a putative neurotrophic factor

FGF-2 presents as an obvious candidate as a regulator of the CNS injury response. It belongs to a family of at least nine homologous factors which are potent effectors of a wide range of cell types (for review see Baird, 1994). The potential of FGF-2 as a neurotrophic factor has been recognized since the localization and characterization of the peptide (Gospodarowicz *et al.*, 1984) and its receptors (Lee *et al.*, 1989) in the CNS. In the normal adult brain, FGF-2 is present in neurons and glia, in the vascular basement membrane, the meninges and in ependymal cells of the ventricular system. Glia and neurons synthesize both the growth factor (Emoto *et al.*, 1989) and its receptor(s) (Wanaka, Johnson & Milbrandt, 1990) throughout the brain and spinal cord, but particularly high levels of expression of FGF-2 are seen in discrete populations of neurons such as the induseum griseum, fasciola cinereum, field CA2 of the hippocampus, the septohippocampal nucleus, cingulate cortex and the subfornical organ. The functional significance of this discrete pattern of expression is not established.

FGF-2 is also found throughout the developing CNS (Gonzalez *et al.*, 1990), where it may be important for the maturation of both cholinergic and dopaminergic neurons. FGF-2 promotes the survival in culture of a wide range of fetal CNS neuronal populations, including cortical, hippocampal, striatal, septal, hypothalamic, mesencephalic and ciliary ganglionic neurons (see Baird, 1994). It enhances outgrowth of their neurites and promotes choline acetyltransferase activity (ChAT) in septal cultures (Grothe & Unsicker, 1989) and dopamine uptake in mesencephalic cultures (Ferrari *et al.*, 1989). In support of a neurotrophic role for this peptide in the developing CNS, exogenous FGF-2 is reported to rescue neurons *in vivo* from normal embryonic degeneration (Dreyer *et al.*, 1989) and also from the photoreceptor degeneration associated with an inherited retinal dystrophy (LaVail *et al.*, 1992).

Despite the early identification of FGF-2 in the normal mature CNS, little is known of its physiological function there. *In vivo* studies do provide some, albeit circumstantial, evidence of a neurotrophic role for endogenous basic FGF in the injured adult CNS. A number of reports have demonstrated increased immunoreactive and bioactive FGF-2 in the chemically (Riva, Gale & Mocchetti, 1991) and mechanically lesioned brain (Finklestein *et al.*, 1988; Frautschy, Walicke & Baird, 1991; Logan *et al.*, 1992*a*), with enhanced expression localized to neurons, glia and vascular endothelial cells within the damaged neuropil. Transient forebrain ischaemia is reported to similarly induce FGF-2 expression in neurons and astrocytes (Takami *et al.*, 1992).

In particular, we have used a model of penetrant brain injury to show that there are rapid and transient local changes in FGF-2 expression and compartmentalization in neurons and glia (Logan *et al.*, 1992*a*). The increase in FGF-2 mRNA and protein peaks between 7 and 10 days post-lesion and thereafter declines to control levels after 14 days. These changes are most dramatically visualized in the population of reactive astrocytes (Fig. 3). It is clear from this that FGF-2 has the potential to act as a paracrine and autocrine neurotrophic factor. However, it does seems from these observations that, whilst increased supplies of FGF-2 are potentially available to responsive neurons and glia in the early wound, its availability is transient, with FGF-2 rapidly becoming limiting at a time co-incident with the abortion of neuronal regeneration.

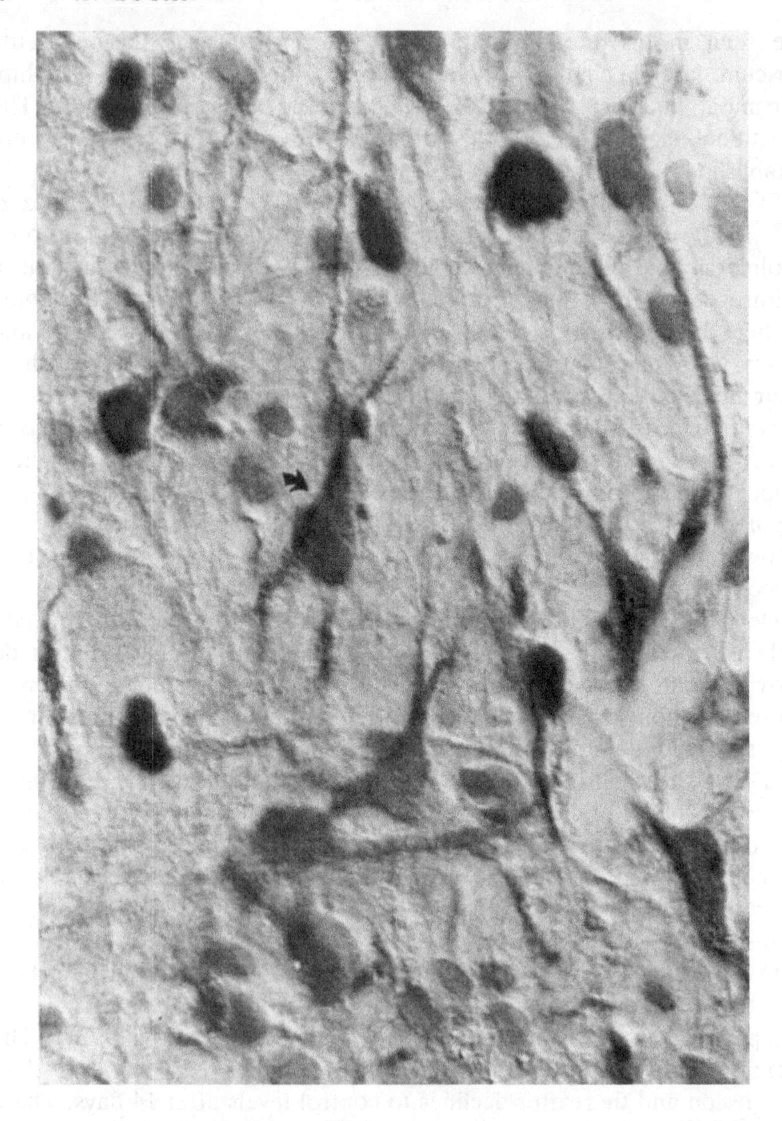

Fig. 3. Immunoperoxidase stained coronal sections through the damaged neuropil of a lesioned adult cerebral cortex. FGF-2 staining is apparent in reactive astrocytes (see solid arrow). Note the presence of both nuclear and cytoplasmic peptide.

In the fetal and perinatal CNS the levels of FGF-2 are constitutively higher both before and after injury (Fig. 4). However, it is clear that in the brain there is a similar rapid increase in FGF-2 mRNA levels in both lesioned and contralateral unlesioned hemispheres (demonstrated by ribonuclease protection assay in Fig. 5). Furthermore, this increase is more sustained than in the adult, mRNA levels still being elevated at 14 days post-lesion and beyond. This suggests that, in the immature brain, FGF-2 may not be a limiting neurotrophic factor, thereby permitting a much more vigorous post-injury neuronal growth response.

Other *in vivo* experiments have provided more indirect evidence of a neurotrophic role of FGF-2 in damaged CNS tissue and indicates its potential as a therapeutic agent to ameliorate neurodegenerative conditions (reviewed by Logan & Berry, 1993). When given exogenously, the peptide promotes the survival of central neurons after chemical insult and axotomy. It also attenuates the decrease of hippocampal ChAT activity induced by partial fimbria transection. In addition recent data has shown that FGF-2 causes a reduction in hippocampal neuronal death after axotomy caused by glutamate, by raising the threshold for glutamate neurotoxicity (Mattson *et al.*, 1989; Skaper, Leon & Facci, 1993). These observations suggest that FGF-2 may play a protective role against the excitotoxic damage that occurs after neuronal insult.

It has been argued that the effects of FGF-2 on neurons is indirect, mediated by glial cells (Engele & Bohn, 1991). However, the identification of functional FGF receptors on neurons infers direct effects, as does the demonstration of receptor–ligand internalization and anterograde/retrograde transportation to neuronal cell bodies, in a manner analogous to nerve growth factor (Walicke & Baird, 1991).

Since FGF-2 has a broad range of activities on multiple cell types *in vitro*, its potential *in vivo* influences in the CNS may spread beyond direct neurotropism (see Baird & Bohlen, 1990). FGF-2 is mitogenic for oligodendrocytes and astrocytes. It stimulates migration and differentiated function of astrocytes, such as release of plasminogen activators and expression of a number of proteins including intermediate filament protein, GFAP, glutamine synthetase and S100 protein. Furthermore, it modifies the morphological maturation of astrocytes, as illustrated by its ability to cause rearrangement of intermediate filaments, to increase extension of cellular processes and to change astrocyte membrane structure. FGF-2 is one of the most potent angiogenic factors so far identified, being both mitogenic and chemotactic for endothelial cells. Its intraventricular administration is reported to promote cerebral angiogenesis in damaged neural tissue after chronic forebrain ischaemia

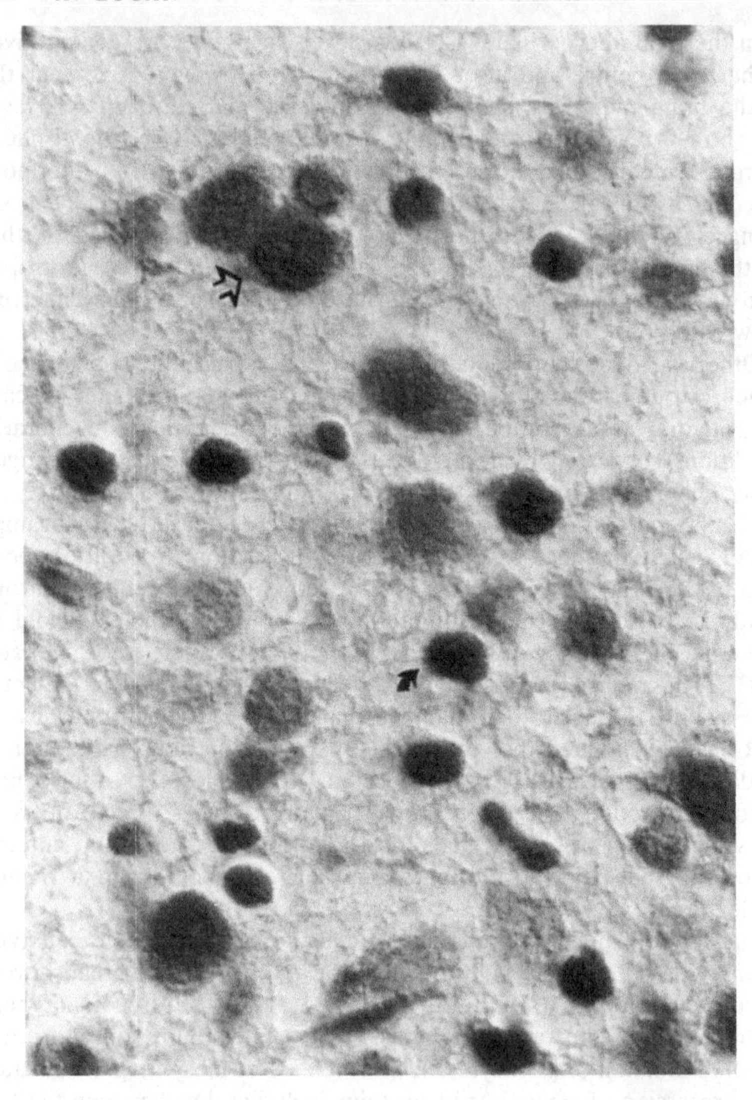

Fig. 4. Coronal sections through the intact cerebral cortex of a neonatal rat. Immunoperoxidase staining reveals FGF-2 peptide localized to both astrocytes (solid arrow) and neurons (open arrow). Note the predominant nuclear staining.

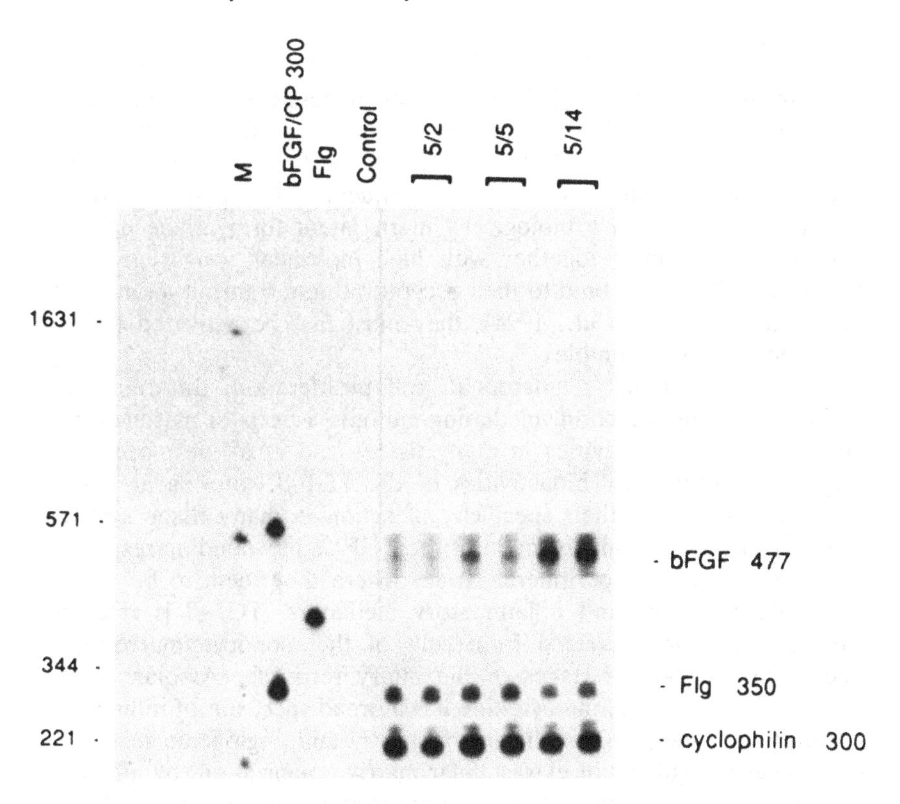

Fig. 5. Identification of FGF-2 and flg receptor mRNA in injured neonatal rat brain by ribonuclease protection assay using total RNA (10 μg/ml) isolated from lesioned ipsilateral and unlesioned contralateral cerebral hemispheres 2, 5 and 14 days post-surgery. Note the increased levels of FGF-2 mRNA in both hemispheres as the wounding response proceeds. For comparison, the relative abundance of cyclophilin mRNA (a house-keeping gene) is shown in the same samples. Lane 1 shows molecular weight markers, lanes 2 and 3 show ^{32}P-labelled FGF-2 (bFGF)/cyclophilin (CP) and flg antisense riboprobes alone. Lane 4 shows the probes after hybridizing with yeast tRNA and RNase treatment.

(Lyons, Anderson & Meyer, 1991). As its name suggests, FGF-2 is also a mitogen and chemoattractant for fibroblasts, promoting the formation of granulation tissue within peripheral wounds *in vivo* (Buntrock *et al.*, 1984). Its relationship to scar formation in the CNS remains to be demonstrated.

TGF-β1 as a putative fibrogenic factor

Five distinct isoforms of TGF-β have been described, although only TGF-β1, -β2 and -β3 are found in mammals (reviewed elsewhere, Roberts & Sporn, 1990). Each isoform is coded for by a separate gene, and each has a distinct 5′ regulatory sequence. The peptides are all secreted from cells in a biologically inert, latent form, made up of a mature TGF-β dimer together with high molecular weight proteins. Before the TGF's can bind to their receptor(s) and transmit an intracellular signal (Wrana *et al.*, 1994), they must first be activated by dissociation from this complex.

TGF-βs are trophic regulators of cell proliferation, differentiation and differentiated function, mediating multiple aspects of mesenchymal and epithelial cell activities in many tissues and at all developmental stages. In culture, the bioactivities of the TGF-β isoforms are often interchangeable, but their specificity of action in many tissue systems *in vivo* is noted. The involvement of the TGF-βs in wounding responses is well established in peripheral tissues where they seem to be potent desmoplastic agents and inflammatory mediators. TGF-β is released from platelets and secreted from cells of the monocyte/macrophage lineage at the earliest stages of the injury response (Assoian *et al.*, 1983). Within wounds, the cytokine has a broad spectrum of influences, not only as a mediator of the inflammatory and angiogenic response, but also as a regulator of extracellular matrix organization, by affecting the expression of many collagens, fibronectins, and elastin, and also modulating the activity of the matrix proteinases and protease inhibitors, which help the remodelling of tissues after damage (reviewed elsewhere, Roberts, McCune & Sporn, 1992). In addition, it is suggested that TGF-β1 controls cell migration and differentiation by influencing the expression of cell adhesion molecules within responding tissues (Heino *et al.*, 1989). Numerous studies have illustrated the potential use of exogenous TGF-β to enhance the repair of bone and skin. Conversely, it is reported that neutralizing antibodies to TGF-β1 can reduce fibrous scar formation in skin with potential cosmetic effects (Shah, Foreman & Ferguson, 1994).

At first site, TGF-β1 does not seem a likely candidate as a CNS trophic factor, since in the mature animal its mRNA is not normally expressed at significant levels by neural tissue, although the peptide has been identified in the meninges and choroid plexus (Logan *et al.*, 1992*b*, 1994). However, evidence is accumulating that TGF-β1 may play a precise role in the CNS, particularly in some pathological conditions.

Studies have shown that astrocytes and neurons which are effectively 'wounded' by their monolayer culture express TGF-β1 mRNA and peptide (Wesselingh *et al.*, 1990). The cytokine is known to stimulate proliferation of astrocytes in primary culture (Johns *et al.*, 1992; Sakai *et al.*, 1990). In addition, TGF-β1 is reported to interact with other trophic factors such as nerve growth factor and FGF-2 to influence multiple aspects of glial function *in vitro*. The peptide affects astrocyte migration and association (Unsicker *et al.*, 1992), and expression of plasminogen activator, fibronectin, collagen, laminin (Toru-Delbauffe *et al.*, 1992), glial fibrillary acidic protein (an intermediate filament protein) (Sakai *et al.*, 1990), interferon-induced MHC class-II antigen (Johns *et al.*, 1992) and cell adhesion molecules such as L1 and N-CAM (Saad *et al.*, 1991). The effects of TGF-β1 on cultured neurons has been less extensively studied but there is some evidence that the peptide can promote neuron survival and neurite outgrowth by modulating the effects of neurotrophic factors like FGF-2 and nerve growth factor (Chalazonitis *et al.*, 1992; Flanders *et al.*, 1991; Lindholm *et al.*, 1992*a*).

More direct evidence of a role for TGF-β1 in CNS wounds has come from *in vivo* experiments. After mechanical lesioning (Nichols *et al.*, 1991; Lindholm *et al.*, 1992*b*; Logan *et al.*, 1992*b*) and hypoxia–ischaemia (Klempt *et al.*, 1992) TGF-β1 is seen to be rapidly and transiently expressed within damaged neural tissue. The cytokine is released from platelets and is also translocated into the wound by invading cells of the monocyte/macrophage lineage. Within hours of injury TGF-β1 is also expressed locally by compromised neurons and glia (Fig. 6) and distally at multiple sites including the meninges, choroid plexus and the endothelial cells of the microvasculature.

Within the damaged tissue TGF-β1 seems to have multiple influences. Some of these have been demonstrated directly by infusing specific anti-TGF-β1 antibodies into CNS wounds to neutralize endogenous TGF-β1 activity (Logan *et al.*, 1994). The consequences of this treatment include a reduction in the number of macrophages and fibroblasts invading the wound and a dramatic inhibition of matrix deposition. Of interest, although a normal reactive gliosis response is observed, the activated astrocytes do not migrate and associate to form a limiting membrane bordering the wound. The suppression of glial-mesodermal scar formation after anti-TGF-β1 treatment demonstrates the primary importance of this cytokine for initiation of CNS scarring. Since it is clear that scar formation in this instance is unhelpful to nerve regeneration, the therapeutic potential of locally applied TGF-β1 antagonists for patients within CNS injuries is implicit.

Since the fetal and perinatal CNS heals without scarring one might predict that a potent fibrogenic factor like TGF-β1 would be absent from such wounds. This is clearly not the case (Fig. 7). The intact immature CNS expresses TGF-β1, -β2 and -β3 at significant levels and

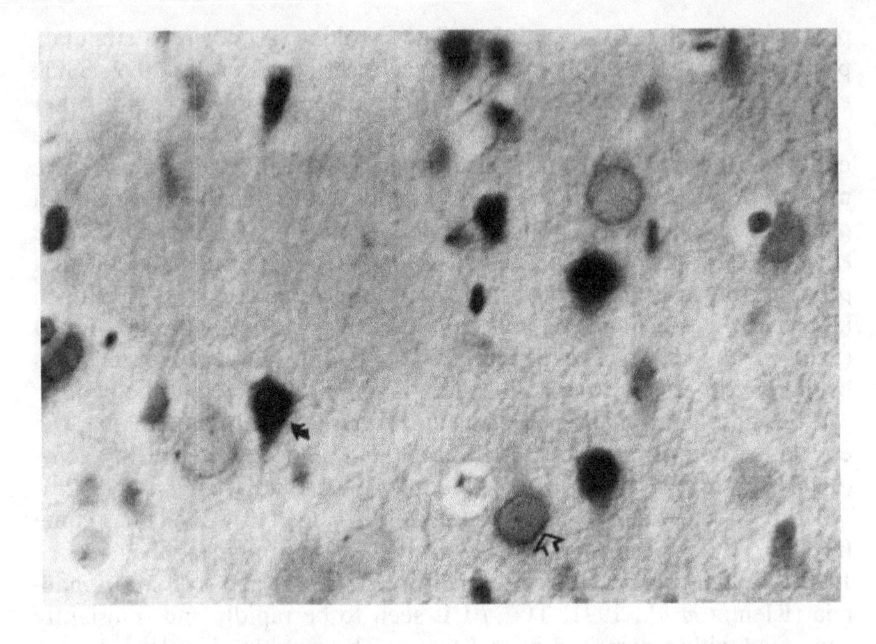

Fig. 6. Immunoperoxidase stained coronal sections through the lesioned cerebral hemisphere of an adult rat, showing TGF-β1 immuno-positive astrocytes (solid arrow) and neurons (open arrow).

a role for this factor in regulation of neuronal migration and differentiation, as well as in glial cell proliferation is suggested (Flanders *et al.*, 1991). Furthermore, there is a rapid increase in TGF-β1 mRNA expression within damaged perinatal brain tissue, visualized by ribonuclease protection assay in Fig. 8. The question of why the immature CNS does not scar remains to be fully answered. However, it may be due to a number of factors relating to the developmental competence of target cells to respond appropriately and the presence or absence of other interactive peptide growth factors. It is clear that individual growth factors do not act in isolation and cellular responses are dependent on the context in which the factor is presented, and this is particularly true of the TGF-βs.

Conclusion

There is much that remains to be done in elucidating the developmental changes that occur in the CNS wounding response. However, it is becoming clear that understanding the biology of wounding in the

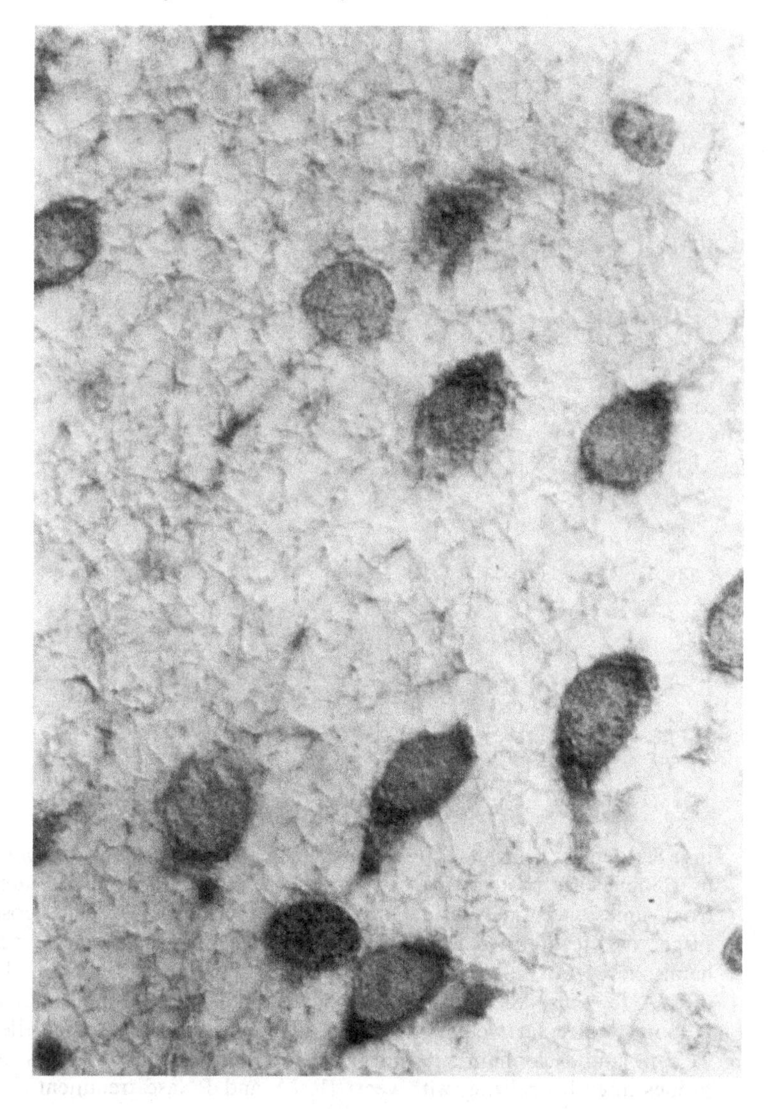

Fig. 7. Immunoperoxidase stained coronal sections through the lesioned cerebral hemisphere of a neonatal rat, showing TGF-β1 immuno-positive astrocytes and neurons.

perinatal CNS, and comparing the trophic and cellular responses to those seen after maturity, may provide us with important clues for developing pharmacological strategies which aid functional recovery of adults with head and spinal cord injuries.

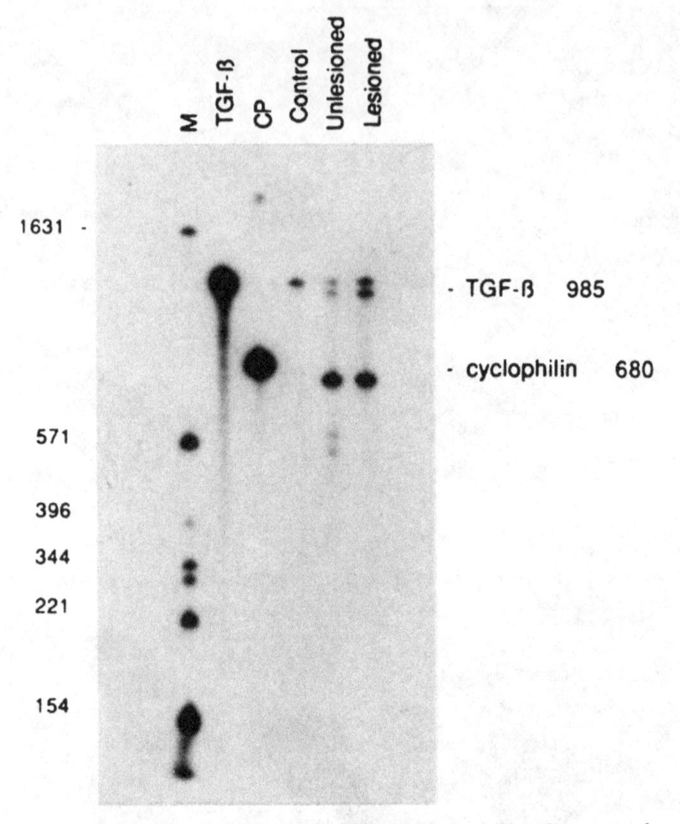

Fig. 8. Identification of TGF-β1 mRNA in injured neonatal rat brain by ribonuclease protection assay using total RNA (10 μg/ml) isolated from unlesioned and lesioned cerebral hemispheres 3 days post-surgery. Note the increased levels of TGF-β1 mRNA in the lesioned hemisphere. For comparison, the relative abundance of cyclophilin mRNA (a house-keeping gene) is shown in the same samples. Lane 1 shows molecular weight markers, lanes 2 and 3 show ^{32}P-labelled TGF-β1 and cyclophilin antisense riboprobes alone. Lane 4 shows the probes after hybridizing with yeast tRNA and RNase treatment.

References

Assoian, R.K., Komoriya, A., Meyers, C.A., Miller, D.M. & Sporn, M.B. (1983). Transforming growth factor beta in human platelets. *Journal of Biological Chemistry*, **258**, 7155–60.

Baird, A. (1994). Fibroblast growth factors: activities and significance of non-neurotrophin neurotrophic growth factors. *Current Opinion in Neurobiology*, **4**, 78–86.

Baird, A. & Bohlen, P. (1990). Fibroblast growth factors. In *Peptide Growth Factors and their Receptors*, ed. M.B. Sporn and A.B. Roberts, pp. 369–418. Vol 95/1. Berlin: Springer-Verlag.

Berry, M., Maxwell, W.L., Logan, A., Mathewson, A., McConnell, P., Ashhurst, D.E. & Thomas, G.H. (1983). Deposition of scar material in the central nervous system. *Acta Neurochirurgica Supplement*, **32**, 31–5.

Buntrock, P., Buntrock, M., Marx, I., Kranz, D., Jentzsch, K.D. & Heder, G. (1984). Stimulation of wound healing, using brain extract with fibroblast growth factor activity. III. Electron microscopy, autoradiography, and ultrastructural autoradiography of granulation tissue. *Experimental Pathology*, **26**, 247–54.

Cajal, S.R.Y. (1928). *Degeneration and Regeneration in the Nervous System*. London: Oxford University Press.

Chalazonitis, A., Kalberg, J., Twardzik, D.R., Morrison, R.S. and Kessler, J.A. (1992). Transforming growth factor β has neurotrophic actions on sensory neurons *in vitro* and is synergistic with nerve growth factor. *Developmental Biology*, **152**, 121–32.

Dreyer, D., Lagrange, A., Grothe, C. & Unsicker, K. (1989). Basic fibroblast growth factor prevents ontogenic neuron death *in vivo*. *Neuroscience Letters*, **99**, 35–8.

Emoto, N., Gonzalez, A.M., Walicke, P.A., Wada, E., Simmons, D.M., Shimasaki, S. & Baird, A. (1989). Identification of specific loci of basic fibroblast growth factor synthesis in the rat brain. *Growth Factors*, **2**, 21–9.

Engele, J. & Bohn, M.C. (1991). The neurotrophic effects of fibroblast growth factors on dopaminergic neurons in vitro are mediated by mesencephalic glia. *Journal of Neuroscience*, **11**, 3070–8.

Ferrari, G., Minozzi, M.C., Toffano, G., Leon, A. & Skaper, S.D. (1989). Basic fibroblast growth factor promotes the survival and development of mesencephalic neurons in culture. *Developmental Biology*, **133**, 140–7.

Finklestein, S.P., Apostolides, P.J., Caday, C.G., Prosser, J., Philips, M.F. & Klagsbrun, M. (1988). Increased basic fibroblast growth factor (bFGF) immunoreactivity at the site of focal brain wounds. *Brain Research*, **460**, 253–9.

Flanders, K.C., Ludecke, G., Engels, S., Cissel, D.S., Roberts, A.B., Kondaiah, P., Lafyatis, R., Sporn, M.B. & Unsicker, K. (1991). Localization and actions of transforming growth factor-β in the embryonic nervous system. *Development*, **113**, 183–91.

Frautschy, S.A., Walicke, P.A. & Baird, A. (1991). Localization of basic fibroblast growth factor and its mRNA after CNS injury. *Brain Research*, **553**, 291–299.

Gonzalez, A.M., Buscaglia, M., Ong, M. & Baird, A. (1990). Distribution of basic fibroblast growth factor in the 18-day rat fetus: localization in the basement membranes of diverse tissues. *Journal of Cell Biology*, **110**, 753–65.

Gospodarowicz, D., Cheng, J., Lui, G.-M., Baird, A. & Bohlen, P. (1984). Isolation of brain fibroblast growth factor by heparin-Sepharose affinity chromatography: Identity with pituitary fibroblast growth factor. *Proceedings of the National Academy of Sciences, USA*, **81**, 6963–7.

Grothe, C. & Unsicker, K. (1989). Basic fibroblast growth factor promotes *in vitro* survival and cholinergic development of rat septal neurons: comparison with the effects of nerve growth factor. *Neuroscience*, **31**, 649–61.

Heino, J., Ignotz, R.A., Hemler, M.E., Crouse, C. & Massague, J. (1989). Regulation of cell adhesion receptors by transforming growth factor-β. *Journal of Biological Chemistry*, **264**, 380–8.

Johns, L.D., Babcock, G., Green, D., Freedman, M., Sriram, S. & Ransohoff, R.M. (1992). Transforming growth factor β1 differentially regulates proliferation and MHC class-II antigen expression in forebrain and brainstem astrocyte primary cultures. *Brain Research*, **585**, 229–36.

Klempt, N.D., Sirimanne, E., Gunn, A.J., Klempt, M., Singh, K., Williams, C. & Gluckman, P.D. (1992). Hypoxia-ischemia induces transforming growth factor β1 mRNA in the infant rat brain. *Molecular Brain Research*, **13**, 93–101.

LaVail, M.M., Unoki, K., Yasumura, D., Matthes, M.T., Yancopoulos, G.D. & Steinbereg, R.H. (1992). Multiple growth factors, cytokines and neurotrophins rescue photoreceptors from the damaging effects of constant light. *Proceedings of the National Academy of Sciences, USA*, **89**, 11249–53.

Lee, P.L., Johnson, D.E., Cousens, L.S., Fried, V.A. & Williams, L.T. (1989). Purification and complementary DNA cloning of a receptor for basic fibroblast growth factor. *Science*, **245**, 57–60.

Lindholm, D., Castren, E., Kiefer, R., Zafra, F. & Thoenen, H. (1992*b*). Transforming growth factor β-1 in the rat brain: increases after injury and inhibition of astrocyte proliferation. *Journal of Cell Biology*, **117**, 395–400.

Lindholm, D., Hengerer, B., Zafra, F. & Thoenen, H. (1992*a*). Transforming growth factor β-1 stimulates expression of nerve growth factor in the rat CNS. *Developmental Neuroscience*, **1**, 9–12.

Logan, A. & Berry, M. (1993). Transforming growth factor β1 and basic fibroblast growth factor in the injured central nervous system: from physiology to pharmacology? *Trends in Pharmacological Sciences*, **14**, 337–43.

Logan, A., Berry, M., Gonzalez, A.M., Frautschy, S.A., Sporn, M.B. & Baird, A. (1994). Effects of transforming growth factor β1 on scar production in the injured central nervous system of the rat. *European Journal of Neuroscience*, **6**, 355–63.

Logan, A., Frautschy, S.A., Gonzalez, A.M. & Baird, A. (1992*a*). A time course for the focal elevation of synthesis of basic fibroblast growth factor and one of its high affinity receptors (flg) following a localized cortical brain injury. *Journal of Neuroscience*, **12**, 3828–37.

Logan, A., Frautschy, S.A., Gonzalez, A.M., Sporn, M.B. & Baird, A. (1992*b*). Enhanced expression of transforming growth factor β1 in the rat brain after a localized cerebral injury. *Brain Research*, **587**, 216–25.

Lyons, M.K., Anderson, R.E. & Meyer, F.B. (1991). Basic fibroblast growth factor promotes *in vivo* cerebral angiogenesis in chronic forebrain ischemia. *Brain Research*, **558**, 315–20.

Mattson, M.P., Murrain, M., Guthrie, P.B. & Kater, S.B. (1989). Fibroblast growth factor and glutamate: opposing roles in the generation and degeneration of hippocampal neuroarchitecture. *Journal of Neuroscience*, **9**, 3728–40.

Maxwell, W.L., Follows, R., Ashhurst, D.E. & Berry, M. (1990*a*). The response of the cerebral hemisphere of the rat to injury. I. The mature rat. *Philosophical Transactions of the Royal Society of London Series B*, **328**, 479–500.

Maxwell, W.L., Follows, R., Ashhurst, D.E. & Berry, M. (1990*b*). The response of the cerebral hemisphere of the rat to injury. II. The neonatal rat. *Philosophical Transactions of the Royal Society of London Series B*, **328**, 501–13.

Nichols, N.R., Laping, N.J., Day, J.R. & Finch, C.E. (1991). Increases in transforming growth factor-β mRNA in hippocampus during response to entorhinal cortex lesions in intact and adrenalectomized rats. *Journal of Neuroscience Research*, **28**, 134–9.

Riva, M.A., Gale, G. & Mocchetti, I. (1992). Basic fibroblast growth factor mRNA increases in specific brain regions following convulsive seizures. *Molecular Brain Research*, **15**, 311–18.

Roberts, A. & Sporn, M.B. (1990). The transforming growth factor-βs. In *Peptide Growth Factors and their Receptors*, ed. M.B. Sporn and A.B. Roberts, pp. 419–472. Vol 95/1. Berlin: Springer-Verlag.

Roberts, A.B., McCune, B.K. & Sporn, M.B. (1992). TGF-β: regulation of extracellular matrix. *Kidney International*, **41**, 557–9.

Saad, B., Constam, D.B., Ortmann, R., Moos, M., Fontana, A. & Schachner, M. (1991). Astrocyte-derived TGF-β2 and NGF differentially regulate neural recognition molecule expression by cultured astocytes. *Journal of Cell Biology*, **115**, 473–84.

Sakai, Y., Rawson, C., Lindburg, K. & Barnes, D. (1990). Serum and transforming growth factor regulate glial fibrillary acidic protein in serum-free-derived mouse embryo cells. *Proceedings of the National Academy of Sciences, USA*, **87**, 8378–82.

Shah, M., Foreman, D.M. & Ferguson, M.W.I. (1994). Neutralizing antibody to transforming growth factor-beta (1,2) reduced cutaneous scarring in adult rodents. *Journal of Cell Science*, **107**, 1137–57.

Skaper, S.D., Leon, A. & Facci, L. (1993). Basic fibroblast growth factor modulates sensitivity of cultured hippocamapal pyramid neurons to glutamate cytotoxicity: Interaction with ganglioside GM1. *Developmental Brain Research*, **71**, 1–8.

Takami, K., Iwane, M., Kiyota, Y., Miyamoto, M., Tsukuda, R. & Shiosaka, S. (1992). Increase of basic fibroblast growth factor immunoreactivity and its mRNA level in rat brain following transient forebrain ischaemia. *Experimental Brain Research*, **90**, 1–10.

Thanos, S. (1991). The relationship of microglial cells to dying neurones during natural neuronal cell death and axotomy-induced degeneration of the rat retina. *European Journal of Neuroscience*, **3**, 1189–207.

Toru-Delbauffe, D., Baghdassarian, D., Both, D., Bernard, R., Rouget, P. & Pierre, M. (1992). Effects of TGF β1 on the proliferation and differentiation of an immortalized astrocyte cell line: Relationship with the extracellular matrix. *Experimental Cell Research*, **202**, 316–25.

Unsicker, K., Engels, S., Hamm, C., Ludecke, G., Meier, C., Renzing, J., Terbrack, H.G. & Flanders, K.C. (1992). Molecular control of neural plasticity by the multifunctional growth factor families of the FGFs and TGF-βs. *Annals of Anatomy*, **174**, 405–7.

Wahl, S.M., Allen, J.B., McCartney-Francis, N., Morganti-Kossmann, T., Ellingsworth, L., Mai, U.E.H., Mergenhagen, S.E. & Orenstein, J.M. (1991). Macrophage- and astrocyte-derived transforming growth factor β as a mediator of central nervous system dysfunction in acquired immune deficiency syndrome. *Journal of Experimental Medicine*, **173**, 981–91.

Walicke, P.A. & Baird, A. (1991). Internalization and processing of basic fibroblast growth factor by neurons and astrocytes. *Journal of Neuroscience*, **11**, 2249–58.

Wanaka, A., Johnson, E.M. Jr. & Milbrandt, J. (1990). Localization of FGF receptor mRNA in the adult rat nervous system by *in situ* hybridization. *Neuron*, **5**, 267–81.

Wesselingh, S.L., Gough, N.M., Finlay-Jones, J.J. & McDonald, P.J. (1990). Detection of cytokine mRNA in astrocyte cultures using the polymerase chain reaction. *Lymphokine Research*, **9**, 177–85.

Wictorin, K., Brundin, P., Sauer, H., Lindvall, O. & Bjorklund, A. (1992). Long distance directed axonal growth from human dopaminergic mesencephalic neuroblasts implanted along the nigrostriatal pathway in 6-hydroxydopamine lesioned adult rays. *Journal of Comparative Neurology*, **323**, 475–94.

Wrana, J.L., Attisano, L., Wieser, R., Ventura, F. & Massague, J. (1994). Mechanism of activation of the TGF-β receptor. *Nature*, **370**, 341–7.

MARION C. DICKSON, JULIE S. MARTIN and ROSEMARY J. AKHURST

The role of transforming growth factors β during cardiovascular development

Introduction

The transforming growth factors β (TGF-βs) are a family of closely related peptides, which act on a variety of cell types, with surprisingly diverse biological actions. In addition to the TGF-βs, first isolated in 1983 (Roberts *et al.*, 1983), there is now an ever-growing super-family containing, the activins and inhibins, and the DVR (*D*ecapentaplegic-*Vg*-*R*elated) group of growth and differentiation factors, which are found in mammals, *Xenopus leavis* and *Drosophila melanogaster*. All of these molecules are thought to be important during development (for review see Akhurst, 1994).

The TGF-β family *per se* has three mammalian isoforms, TGF-β1, TGF-β2 and TGF-β3, which are synthesized as precursor molecules and later cleaved to form an amino terminal latency-associated peptide and a mature carboxy terminal protein of 112 amino acids. In the biologically active region there is approximately 90% amino acid sequence homology between the three isoforms, suggesting that the different molecules should display similar biological activities (Roberts & Sporn, 1990). However, although the three isoforms have similar qualitative and quantitative effects on keratinocytes, fibroblasts, and osteoclasts, this is not always the case. TGF-β1 has been shown to be a potent inhibitor of endothelial cells *in vitro* (Heimark, Twardzik & Schwartz, 1986; Muller *et al.*, 1987), whereas TGF-β2 is one to two orders of magnitude less effective as an endothelial growth inhibitor (Jennings *et al.*, 1988; Merwin *et al.*, 1991; Qian *et al.*, 1992). Similarly, TGF-β2 can be an inducer of mesoderm formation in *Xenopus* ectodermal explants, whereas TGF-β1 only possesses the ability to potentiate the mesoderm-inducing capacity of basic fibroblast growth factor (Rosa *et al.*, 1988).

The reduced biological activity of TGF-β2 on endothelial cells (Qian *et al.*, 1992) is probably due to the differential affinities of binding of the different TGFβ isoforms to the recently cloned serine/threonine kinase TGF-β type I and II receptors (TβRI and TβII) (Lin *et al.*, 1992; Franzen *et al.*, 1993; Wrana *et al.*, 1994). TβRII is thought to be the major receptor through which TGF-βs bind to the hetero-oligomeric receptor complex (Fig. 1) Ligand binding to TβRII activates TβRI, by phosphorylation, and thus triggers the cascade of intracellular signalling events which results in a biological response (Fig. 1. Wrana *et al.*, 1994). TGF-β2 has low affinity for TβRII, and requires a third cell surface protein, β-glycan or the type III receptor, to deliver TGF-β2 effectively to TβRII (Lin *et al.*, 1992; Henis *et al.*, 1994). In contrast, TGF-β1 and TGF-β3, which have a spectrum of quantitative biological responses more similar to each other than to those of TGF-β2 (Graycar *et al.*, 1989; Cheifetz *et al.*, 1990), both bind to TβRII with high affinity (Lin *et al.*, 1992). Interestingly, β-glycan is not expressed on endothelial or haematopoietic cells. Instead, a proteoglycan related to β-glycan,

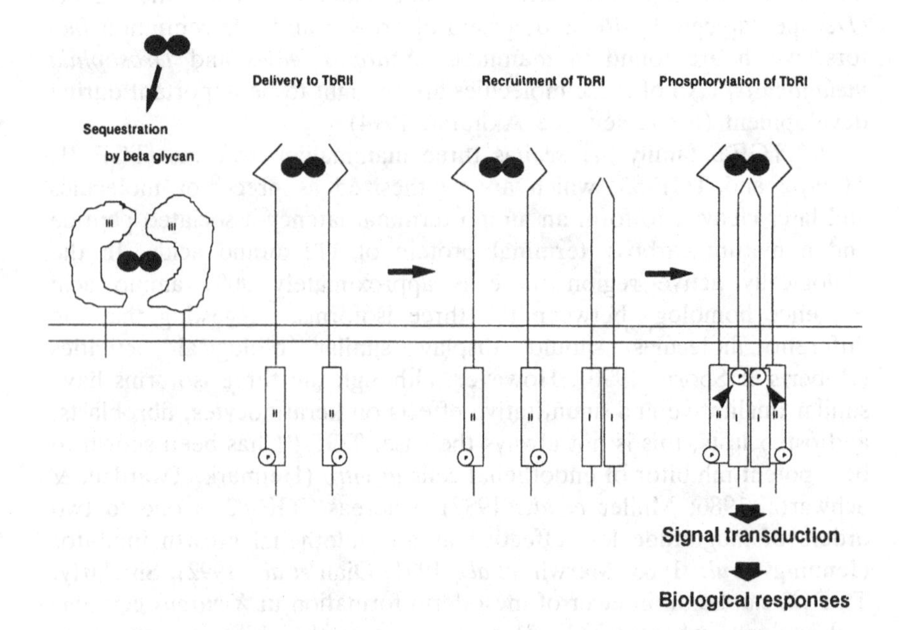

Fig. 1. TGF-β receptor binding and signalling by dimeric TGF-βs. Cartoon to indicate the current consensus on the mechanism of binding and activation of the TGF-β receptors by their ligands, based on models by Henis *et al.* (1994) and Wrana *et al.* (1994). Blocked-in dumbell denotes TGF-β ligand; I, TβRI; II, TβRII; III, betaglycan or TβRIII; P denoted state of phosphorylation of receptors I and II.

termed endoglin, is expressed on endothelial cells (Cheifetz *et al.*, 1992). Endoglin, unlike β-glycan, has low affinity for TGF-β2, possibly explaining the low activity of this isoform on endothelial cells.

Several approaches have been taken to address the exact role of the TGF-β family in mammalian fetal development. This includes the wealth of information gathered on the *in vitro* biological activities of the TGF-βs (Roberts & Sporn, 1990; Akhurst, 1994), descriptive studies on the spatial and temporal sites of expression of the TGF-β genes and their encoded proteins during mammalian development (Akhurst *et al.*, 1992) and, more recently, the generation of transgenic mice with specific ablation of individual TGF-β genes (Shull *et al.*, 1992; Kulkarni *et al.*, 1993).

TGF-βs 1 and 2 are implicated in cardiovascular development

The early mammalian heart tube is composed essentially of two cell types, an inner endothelial tube (the endocardium) and an outer myocardial tube. The development of this early heart tube therefore involves two major differentiative processes; a) differentiation of endothelial/endocardial cells from mesoderm, termed vasculogenesis, b) differentiation of cardiomyocytes from cardiac mesoderm, or cardiomyogenesis. Both of these cell types arise from the splanchnic mesoderm, which is located ventrally within the body cavity of the 7.5 to 8.0 day *post-coitum* (p.c.) mouse embryo. *In situ* hybridization and immunolocalization studies have implicated TGF-β1 (Akhurst *et al.*, 1990) and TGF-β2 (Dickson *et al.*, 1993) in vasculogenesis and cardiomyogenesis, respectively.

The progression of the heart from a simple tube, to a complex four-chambered organ, involves massive growth, complex morphogenetic movements, and tissue interactions that result in the development of a third tissue type, the cardiac cushion mesenchyme. This appears, interposed between the endocardium and myocardium, and contributes to formation of valves and septae (Potts *et al.*, 1991). Descriptive studies have also implicated TGF-β1 and TGF-β2 in cardiac morphogenesis and cardiac cushion tissue formation, respectively (Heine *et al.*, 1987; Akhurst *et al.*, 1990; Millan *et al.*, 1991; Dickson *et al.*, 1993). TGF-β3 does not appear to be expressed in mouse cardiac development until later stages, when expression is first seen in the pericardium and in the established cardiac mesenchyme of the heart valves (Millan *et al.*, 1991), although in the chick, it has been implicated in the earlier event of cardiac cushion tissue induction (Potts *et al.*, 1991).

TGF-β2 expression: a role in cardiomyogenesis?

The cardiac mesoderm, or pro-myocardium, constitutes a crescent-shaped region of splanchnic mesoderm at the rostral end of the embryo, which has the potential to differentiate into myocardium (Rosenquist & de Haan, 1966). The cardiac mesoderm is thought to be induced by factors secreted from the underlying foregut endoderm (Muslin & Williams, 1991). Cardiomyogenesis proceeds in a rostro-caudal, ventriculo-atrial direction, as pro-myocardial mesoderm cells migrate into the elongating heart tube at the caudal end (that is, via the vitelline veins and sinus venosus), and differentiate into cardiomyocytes.

Detailed *in situ* and immunohistochemical studies on 7.0–9.5 dpc embryos have shown that TGF-β2 is first expressed in the crescent-shaped pro-myocardium at the 1-2 somite stage (Dickson *et al.*, 1993). By the 5-7 somite stage, intense TGF-β2 RNA expression is detected in the caudal splanchnic mesoderm lining the intraembryonic coeloma (Fig. 2), which is destined to migrate into the cardiac tube and contribute to myocardium (Rosenquist & de Haan, 1966). In close association with this, the pro-myocardium of the vitelline veins also expresses TGF-β2 transcripts, as does the sinus venosus, albeit at lower levels (Fig. 2). Thus there is a gradient of TGF-β2 gene expression, with highest levels in the undifferentiated mesoderm at the caudal, inflow end of the heart, and reduced levels in differentiating cardiomyocytes towards the rostral outflow end. Interestingly, the tissues surrounding the pericardial cavity, including the splanchnic and somatic epithelia, ventral foregut endoderm and budding thyroid diverticulum also showed intense expression of TGF-β2 (Fig. 2).

By 8.5 dpc the heart tube is undergoing a series of morphological changes brought about by differential growth rates along the heart tube, whilst under the constraints of a small pericardial cavity. This causes the heart tube to bend and dilate, contributing to formation of the four chambers of the heart. During this stage of intense morphogenesis, TGF-β2 expression is maintained in the regions of the splanchnic mesoderm feeding into the inflow and, to a lesser extent, outflow tracts of the heart, and in the ventral foregut endoderm, lying next to the heart. However, very low levels of TGF-β2 RNA expression were seen in the differentiated myocardium of the heart *per se* (Dickson *et al.*, 1993).

Using immunohistochemical techniques TGF-β2 protein was first detectable at 8.5 dpc, in the differentiated cardiomyocytes of the bulbus cordis, ventricle and atrium (Dickson *et al.*, 1993). Paradoxically, the protein staining pattern was directly opposed to that of the RNA

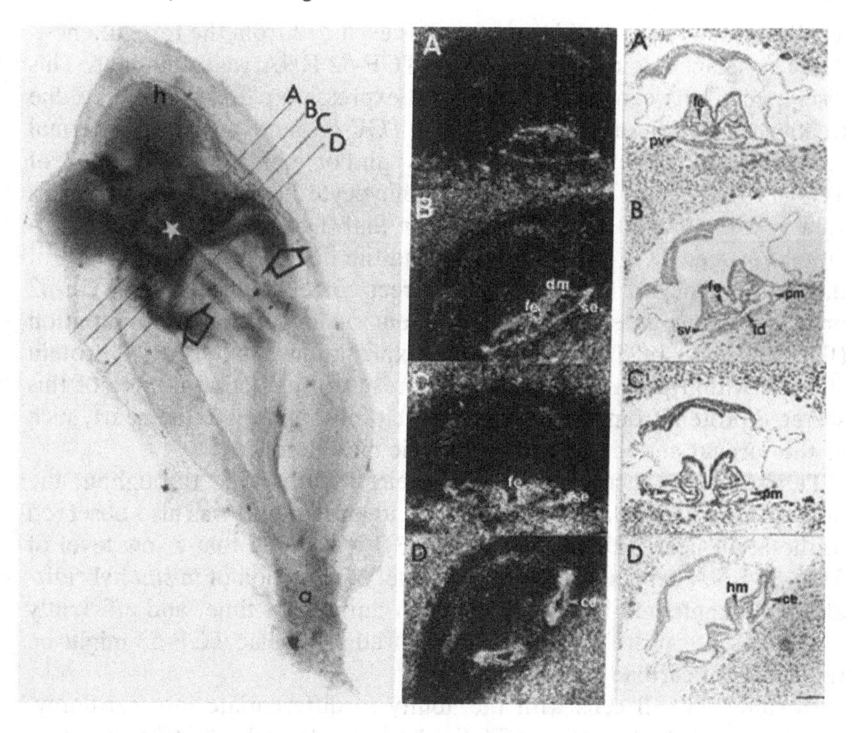

Fig. 2. Expression of TGF-β2 gene in early cardiac development. Left panel shows whole mount *in situ* hybridization of a 8.0 dpc mouse embryo with a cRNA TGF-β2 probe, to show expression (dark stain) in the splanchnic mesoderm feeding into the heart. Arrows indicate the direction of migration of mesoderm cells into the myocardium (Rosenquist & deHaan, 1966). a, allantois; h, head; *, heart. Lines labelled A–D, denote the level of the sections in the adjacent photomicrographs of radioactive sections *in situ* hybridizations. Middle panels (A–D) are dark field (signal = white); right panels (A–D) are bright field images of *in situ* hybridization with the TGF-β2 probe to show cellular detail of hybridization pattern, and also to emphasize the rostro-caudal gradient of TGF-β2 gene expression. ce, coelomic epithelium; dm, dorsal mesocardium; fe, foregut endoderm; hm, head mesenchyme; pm, promyocardium; pv, primitive ventricle; se, splanchnic epithelium; sv, sinus venosus; td, thyroid diverticulum; vv, vitelline vein. Photographs taken from Dickson *et al.* (1993).

localization, as polypeptide staining was excluded from the foregut endoderm or splanchnic mesoderm where TGF-β2 RNA was abundant. This discrepancy between protein and RNA expression patterns could be due to downregulation of expression of the TGF-β2 gene, as the mesodermal cells differentiate into cardiomyocytes, and/or inefficient translation of TGF-β2 RNA until the mature cardiomyocyte has differentiated. This view is strengthened by the observation that there was an inverse correlation between TGF-β2 RNA levels and the degree of differentiation of the cardiomyocyte, compared to a direct correlation between TGF-β2 protein staining intensity and the extent of myocardial differentiation (Dickson *et al.*, 1993). An alternative explanation is that TGF-β2 protein staining within the myocardium is the result of paracrine uptake of this secreted protein from synthetic sources closely apposed to the heart, such as the foregut endoderm and splanchnic mesoderm.

Persistent presence of TGF-β2 protein was observed throughout the myocardium during later stages of organogenesis, and was also observed in the adult heart (Dickson *et al.*, 1993). It is likely that a low level of TGF-β2 RNA expression, below the level of detection of *in situ* hybridization, is maintained in the myocardium during this time, and efficiently translated into protein. The function of adult cardiac TGF-β2 might be regulation of cardiac beat rate.

In summary, all cells with the ability to differentiate into cardiomyocytes express high levels of TGF-β2 RNA although, as differentiation proceeds, RNA expression is down regulated with a concomitant increase in protein levels. TGF-β has generally been thought to be an inhibitor of skeletal myogenesis (Olsen *et al.*, 1986; Massague *et al.*, 1986), acting indirectly via effects on the extracellular matrix (Heino & Massague 1990) and directly via down-regulation of myogenic factors (Vaidya *et al.*, 1989; Heino & Massague, 1990). However, more recently, several groups, have suggested that TGF-β may actually be an inducer of both cardiac and skeletal myogenesis. In support of this, Slager *et al.* (1993) illustrated that both TGF-β1 and TGF-β2 are capable of initiating cultured embryonic stem cells to differentiate into both cardiac and skeletal muscle cells. TGF-β1 has also been implicated in the induction of cardiomyogenesis by pharyngeal endoderm, in the *axolotl* (Muslin & Williams, 1991). Finally, Zentella and Massague (1992) showed that TGF-β, when added to the skeletal myoblast cell line, L_6E_9, cultured under the proliferation-favouring conditions of serum-rich medium, could induce terminal differentiation, as assayed by the up-regulation of muscle-determining genes. The function of the very high levels of expression of TGF-β2 seen in association with development of the early heart tube (Fig. 2), might therefore be an inducer, or modulator, of cardiomyogenesis. More recently, it has been shown that TGF-βs can regulate the rate of cardiomyocyte beat in culture (Roberts *et al.*, 1992), and an endogenous role as an early physiological regulator of cardiac beat is equally plausible.

TGF-β1 expression: a role in vasculogenesis and angiogenesis?

Concomitant with the development of the pro-myocardial plate at the anterior end of the embryo, certain splanchnic mesoderm cells, located immediately beneath the pro-myocardium, undergo vasculogenesis, differentiating first into angioblasts and then endothelial cells, to form the endocardial tube which eventually becomes engulfed by the myocardium at around 8.0 dpc (De Ruiter *et al.*, 1992). The factors which induce vasculogenesis are as yet undetermined.

At the same time that vasculogenesis occurs within the embryo proper, focal regions of mesoderm within the extraembryonic yolk sac also undergo this process, as individual cells differentiate into angioblasts and contribute to the development of the vascular network of the yolk sac. These 'blood islands', in addition to differentiating into endothelial cells, contain the mesodermal progenitors of the early haematopoietic system. TGF-β1 is expressed in both the endothelial and haematopoietic lineages (Lehnert & Akhurst, 1988; Akhurst *et al.*, 1990).

At 7.0 dpc the TGF-β1 gene is expressed at both embryonic and extraembryonic sites of vasculogenesis (Akhurst *et al.*, 1990). As cardiovascular development proceeds, TGF-β1 RNA expression continues within the cells of the endocardium and the major blood vessels, and is also seen in endothelial cells in regions of active angiogenesis, such as the capillary network within the head mesenchyme and the plexus of blood vessels forming around the neural tube. After about 9.5 dpc, TGF-β1 gene expression within the heart persists only in those regions of the endocardium actively undergoing morphogenesis, such as the endocardium overlying the cushion tissue (10.5 dpc), and the valves (to 7 days *post-partum*). Thus, prongioblasts, immature and morphogenetically active endothelial cells all express TGF-β1 and the gene is down-regulated once differentiation and tissue remodelling is complete. This pattern of expression has implicated TGF-β1 as playing a role in vasculogenesis, angiogenesis, and cardiac morphogenesis (Akhurst *et al.*, 1990). The only puzzle has been the fact that TGF-β1, a potent inhibitor of endothelial cell growth (Heimark *et al.*, 1986; Muller *et al.*, 1987), should be expressed in endothelial cells at a time of most rapid endothelial cell proliferation. Hypotheses to explain this are that

TGF-β1 is expressed to modulate endothelial proliferation, or that the response of primitive endothelial cells to TGF-β1 at early embryonic stages, rather than being anti-proliferative, is stimulation of differentiation (Madri, Pratt & Tucker, 1988; Madri, Bell & Merwin, 1992).

Gene targeting demonstrates that TGF-β1 is essential for vasculogenesis

The powerful tool of gene targeting by homologous recombination in embryonal stem cells enables the generation of mice with specified non-functional genes (Mansour, Thomas & Capecchi, 1988). To address the role of TGF-β1 in pre-natal and adult life, two independent laboratories generated lines of mice with one allele of the TGF-β1 rendered inactive by homologous recombination (Shull *et al.*, 1992; Kulkarni *et al.*, 1993). When heterozygous TGF-β1 adults were crossed, developmentally normal homozygous TGF-β1 null pups were born, though the homozygous TGF-β1 null animals could not be kept as a line since they died, at around three weeks of age, from a multifocal inflammatory wasting syndrome (Shull *et al.*, 1992; Kulkarni *et al.*, 1993). In the light of studies on expression of TGF-β1 in the cardiovascular system (Akhurst *et al.*, 1990), and elsewhere (Lehnert & Akhurst, 1988), the normal appearance of the TGF-β1 homozygous null mice might seem a surprising observation. However, later studies showed that when some null pups underwent gestation in an exceptional homozygous TGF-β1 null mother, they *did* have severe cardiac abnormalities, in the form of reduced ventricular lumina and disorganised ventricular muscle and valves, dying shortly after birth as a consequence (Letterio *et al.*, 1994). Letterio *et al.* (1994) showed that transfer of TGF-β1 from mother to fetus can occur via the placenta, and might be responsible for rescuing the null pups from this developmental fate in heterozygous mothers (Letterio *et al.*, 1994). This would explain the high incidence of cardiac defects in the four out of four homozygous TGF-β1 null pups born to the homozygous null mother.

Although some phenotypically normal homozygous TGF-β1 null pups are born to heterozygote crosses, the genotype ratios of wild type: heterozygote: homozygote null are skewed, suggesting that around half of the homozygous TGF-β1 null animals are lost, either before or at birth (Shull *et al.*, 1992; Kulkarni *et al.*, 1993). Thus, TGF-β1 null mice can be classified into two separate categories; those that die pre-natally and those that survive until weaning. Pre-natal lethality of at least some homozygous TGF-β1 animals is more consistent with this growth factor playing a crucial role in development.

To examine the stage at which pre-natal lethality of the TGF-β1 homozygous nulls occurred, genotype analysis of conceptuses from 8.5–18.5 dpc and of pups at 14 days *post partum* (dpp) was performed using the polymerase chain reaction (PCR) technique (Dickson *et al.*, 1995). The numbers obtained from this analysis confirmed that there was a statistically significant deviation from the normal Mendelian ratio of 1 : 2 : 1 (+/+ : +/− : −/−) to 1 : 1.6 : 0.5 at 14 dpp. This same genotype distribution was seen even before birth, at 11.5 and 18.5 dpc, demonstrating that perinatal loss of homozygous null pups had not occurred. Normal 1 : 2 : 1 genotype ratios were seen at 9.5 dpc, suggesting that embryo loss occured between 9.5 and 11.5 dpc.

To determine the cause of pre-natal lethality, detailed descriptive analysis was performed on genotyped 8.5, 9.5 and 10.5 dpc embryos from TGF-β1 heterozygous crosses. At 8.5 dpc, there were no obvious phenotypic abnormalities. However, by 9.5 dpc, approximately 50% of the TGF-β1 null and 20% of the heterozygous conceptuses showed specific defects in the yolk sac. These defects occured in the vascularization of the yolk sac, and ranged from a decrease in vessel number to disorganised and delicate vessels. Occasionally, no chorioallantoic connection was formed. In addition to this, many of the TGF-β1 null yolk sacs were overtly anaemic. The defects were entirely restricted to extraembryonic tissues, the embryos themselves had a normal vasculature and developing haematopoietic system. The TGF-β1 null embryos dissected from defective yolk sacs were, however, developmentally retarded by about 0.5 dpc, and in some cases by up to 2 days. They were often oedemic and/or necrotic, which was presumably a consequence of poor nutrition and oxygenation, normally provided via the chorioallantoic placental circulation (Dickson *et al.*, 1995).

Since Letterio *et al.* (1994) had proposed that maternal transfer of TGF-β1 may rescue the 50% of null animals which survive to birth, an *in vitro* whole embryo culture system was developed in a medium containing minimal levels of TGF-β1 (<0.1 ng/ml) (Martin & Akhurst, unpublished observations). 8.5 dpc embryos, from heterozygote intercrosses were cultured for 24 hrs in this TGF-β1 -depleted medium, with the expectation that this might increase the percentage of homozygous null embryos manifesting the abnormal phenotype. Surprisingly, the number of null embryos which showed defects in the vascularization of the yolk sacs was not amplified in culture, implying that either (1) rescue by maternal TGF-β1 had occurred prior to 8.5 dpc, (2) that rescue of the null embryos might be by other fetal TGF-β isoforms (e.g. TGF-β (3), rather than by maternal TGF-β1 and/or (3) that other genetic factors, which influence TGF-β metabolism, determine the outcome of the phenotype.

In several lines of mice with other genes 'knocked out', it has now been shown that the penetrance of the phenotype is dependent on the genetic background of the mice (Liu *et al.*, 1993; George *et al.*, 1993), indicating that modifier genes may play a vital role in determining phenotypic outcome. Indeed, data generated from the TGF-β1 knock-out mice do suggest that genetic background influences the phenotype (H. Su and C. Biron pers. comm). The most likely candidates for modifier genes in the TGF-β1 knock out system are those which normally control the activation of TGF-β1, such as the insulin-like growth factor II/mannose-6 phosphate receptor (Dennis & Rifkin, 1991), or the latent TGF-β1 binding protein (Flaumenhaft *et al.*, 1993). Other possibilities are proteins which are involved in modulating binding of TGF-β to its receptor, such as β-glycan, or proteins of the receptor/signal transduction pathway.

TGF-β1 has been shown to be a potent negative regulator of endothelial cell proliferation (Heimark *et al.*, 1986; Muller *et al.*, 1987), as well as having effects on endothelial cell differentiation (Madri *et al.*, 1988, 1992). Examination of dissected yolk sacs indicated no excessive proliferation of endothelial cells in homozygous TGF-β1 null material, but there was disruption of cellular adhesion between endothelial layers of the blood vessels in null yolk sacs compared to wild type. In some cases, this had lead to rupture of the endothelial tubes, allowing blood to spill into the yolk sac cavity. It therefore appears that, the genetic lesion resulted in abnormalities in regulation of differentiation rather than proliferation (see Fig. 3). To investigate this further, and to establish the onset of this developmental defect, molecular analysis was performed on yolk sacs prior to the appearance of a morphological phenotype. whole-mount *in situ* hybridization, using a marker for endothelial differentiation, *flk1* (Yamaguchi *et al.*, 1993), was performed on 8.5 dpc conceptuses. Interestingly, the majority of TGF-β1 null yolk sacs showed significantly less hybridization with the *flk1* probe than their wild type litter mates, indicating that the primary cause of the null phenotype was indeed delayed or defective differentiation of the endothelial cell lineage (Dickson *et al.*, 1995, Fig. 3).

In addition to effects on vascular development, by 9.5 dpc the genetic lesion sometimes also resulted in yolk sac anaemia. Whether the anaemia was due to lack of haemoglobinization or a reduction in erythroid cell number was addressed by analysing embryos and yolk sacs after whole-mount *in situ* hybridization with a ζ-globin probe (Wilkinson *et al.*, 1987). Over 50% of the 9.5 dpc TGF-β1 null embryos showed irregular or no ζ-globin staining on the anti-mesometrial side of the yolk sac, with no obvious blood cells within any vessels in

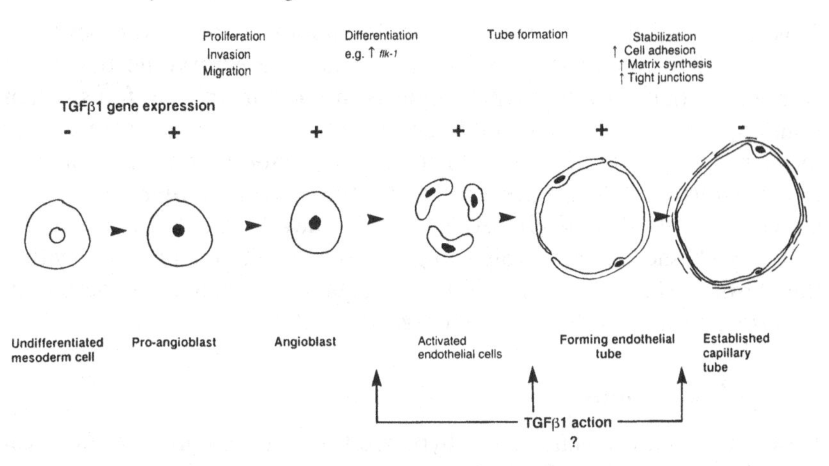

Fig. 3. TGF-β1 expression and function in yolk sac vasculogenesis. Highly schematized cartoon to summarize the temporal expression of TGF-β1 during vasculogenesis, and the possible events controlled by TGF-β1. + or − denote expression as seen by *in situ* hybridization (Akhurst *et al.*, 1990). It should be emphasized, that endothelial differentiation involves a sequence of events, the earliest probably being *flk-1* expression, and the latest probably elaboration of the extracellular matrix, which cannot be adequately portrayed in this cartoon.

comparison to wild type litter mates. The overall erythroid cell number was reduced by approximately 90% in the anaemic yolk sacs, compared to wild types. However, the percentage of total blood cells hybridizing to ζ-globin (i.e. haemoglobinized) was similar in nulls and wild types (Dickson *et al.*, 1995). Thus, although TGF-β has been reported to be a potent growth inhibitor of haematopoietic cells (Ohta *et al.*, 1987; Sing *et al.*, 1988, 1989; Ottman & Pelus, 1988, Keller *et al.*, 1988), the predominant effect of ablation of this growth regulator is not excessive haematopoietic proliferation, but probably defective differentiation (Chen *et al.*, 1989).

It was interesting that the phenotypic defects of TGF-β1 homozygous null conceptuses seemed to be restricted to extra-embryonic tissues. The vascular and haematopoietic systems of the null embryos themselves appeared to be developing normally when they died, as an indirect consequence of the yolk sac defects. It would therefore appear that development of the embryonic and extraembryonic vascular and haematopoietic systems are under different sets of molecular controls.

It is also possible that extraembryonic endothelia are more sensitive to reduced threshold levels of TGF-β1 than are embryonic tissues. It is notable that developmental defects *do* occur in the fetus when homozygous null animals undergo gestation in a homozygous null mother (Letterio *et al.*, 1994). The cardiac abnormalities seen in such homozygous null pups (Letterio *et al.*, 1994), could either be a direct effect of reduced TGF-β1 levels on fetal development, or secondary to the yolk sac defects, which might result in altered pressure within the cardiovascular system, with consequent ventricular hyperplasia/hypertrophy and occlusion (Heine *et al.*, 1985).

Conclusions

Descriptive studies using *in situ* hybridization and immunohistochemical analysis implicated TGF-β1 as playing a role in vasculogenesis, angiogenesis and/or cardiac valve morphogenesis (Akhurst *et al.*, 1990). Studies on TGF-β1 knock-out mice have now confirmed that TGF-β1 is essential for vasculogenesis, at least in extraembryonic tissues (Dickson *et al.*, 1995), and that TGF-β1 knock-out mice which underwent gestation in a homozygous TGF-β1 null mother had cardiac abnormalities (Letterio *et al.*, 1994). Ablation of TGF-β1 *in vivo*, in the knock-out mice, results in defective cellular differentiation rather than cellular proliferation.

Descriptive studies on TGF-β2 expression have implicated this isoform in cardiomyogenesis or cardiomyocyte funtion (Dickson *et al.*, 1993; Roberts *et al.*, 1992), and/or in induction of cardiac cushion tissue (Millan *et al.*, 1991). As yet, neither heterozygous nor homozygous TGF-β2 knock-out mice have been successfully generated. Confirmation of these implications therefore remains to be demonstrated.

Acknowledgements

Work in the laboratory is supported by the MRC, Wellcome Trust and CRC.

References

Akhurst, R.J., Lehnert, S.A., Faissner, A.J. & Duffie, E. (1990). TGF-β in murine morphogenetic processes: the early embryo and cardiogenesis. *Development*, **108**, 645–56.

Akhurst, R.J., FitzPatrick, D.R., Fowlis, D.J., Gatherer, D., Millan, F.A. & Slager, H. (1992). The role of TGF-βs in development and neoplasia. *Molecular Reproductive Development*, **32**, 127–35.

Akhurst, R.J. (1994). The transforming growth factor β family in vertebrate embryogenesis. In *Growth Factors and Signal Transduction in Development*, ed. Nilsen-Hamilton, M. pp. 97–122. New York: Wiley-Liss.

Cheifetz, S., Bellon, T., Cales, C., Vera, S., Bernabeu, C., Massague, J. & Letarte, M. (1992). Endoglin is a component of the transforming growth factor-β receptor system in human endothelial cells. *Journal of Biological Chemistry*, **267**, 19027–30.

Cheifetz, S., Hernandez, H., Laiho, M., ten Dijke, P., Iwata, K.K. & Massague, J. (1990). Distinct TGF-β receptor subsets as determinants of cellular responsiveness to three TGF-β isoforms. *Journal of Biological Chemistry*, **265**, 20533–8.

Chen, L.L., Dean, A., Jenkinson, T. & Mendelsohn, J. (1989). Effect of transforming growth factor-β1 on proliferation and induction of hemoglobin accumulation in K-562 cells. *Blood*, **74**, 2368–75.

Dennis, P.A. & Rifkin, D.B. (1991). Cellular activation of latent transforming growth factor β requires binding to the cation-independent mannose 6-phosphate/insulin-like growth factor receptor. *Proceedings of the National Academy of Sciences, USA*, **88**, 580–4.

DeRuiter, M.C., Poelmann, R.E., VanderPlas-de Vries, I., Mentink, M.M.T. & Gittenberger-de Groot, A.C. (1992). The development of the myocardium and endocardium in mouse embryos: fusion of two heart tubes? *Anatomical Embryology*, **185**, 461–73.

Dickson, M.C., Slager, H.G., Duffie, E., Mummery, C.L. & Akhurst, R.J. (1993). TGF-β2 RNA and protein localizations in the early embryo suggest a role in cardiac development. *Development*, **117**, 625–39.

Dickson, M.C., Martin, J.S., Cousins, F.M., Karlssonn, S., Kulkarni, A.B. & Ackhurst, R.J. (1995). Transforming growth factor β is essential for haematopoiesis and endothelial differentiation *in vivo*. *Development*, **121**, 1845–54.

Flaumenhaft, R., Abe, M., Sato, Y., Miyazono, K., Harpel, J., Heldin, C. & Rifkin, D.B. (1993). Role of latent TGF-β binding protein in the activation of latent TGF-β by co-cultures of endothelial and smooth muscle cells. *Journal of Cell Biology*, **120**, 995–1002.

Franzen, P., ten Dijke, P., Ichijo, H., Yamashita, H., Schultz, P., Heldin, C. & Miyazono, K. (1993). Cloning of a TGF-β type I receptor that forms a heterodimeric complex with the TGF-β type II receptor. *Cell*, **75**, 681–92.

George, E.L., Georges-Labouesse, E.N., Patel-King, R.S., Rayburn, H. & Hynes, R.O. (1993). Defects in mesoderm, neural tube and vascular development in mouse embryos lacking fibronectin. *Development*, **119**, 1079–91.

Graycar, J.I., Miller, D.A., Arrick, B.A., Lyons, R.M., Moses, H.L. & Derynck, R. (1989). Human TGF-β3: Recombinant expression, purification and biological activities, compared with TGF-β1 and TGF-β2. *Molecular Endocrinology*, **3**, 1977–86.

Heimark, R.L., Twardzik, D.R. & Schwartz, S.M. (1986). Inhibitions of endothelial regeneration by type-β transforming growth factor from platelets. *Science*, **233**, 1078–80.

Heine, U.I., Munoz, E.F., Flanders, K.C., Ellingsworth, L.R., Lam, H.Y., Thompson, N.L., Roberts, A.B. & Sporn, M.B. (1987). Role of transforming growth factor-β in the development of the mouse embryo. *Journal of Cell Biology*, **105**, 2861–76.

Heine, U.I., Roberts, A.B., Munoz, E.F., Roche, N.S. & Sporn, M.B. (1985). Effects of retinoid deficiency on the development of the heart and vascular system of the quail embryo. *Virchows Archive (Cell Pathology)*, **50**, 135–52.

Heino, J. & Massague, J. (1990). Cell adhesion and decreased myogenic gene expression implicated in the control of myogenesis by transforming growth factor-β. *Journal of Biological Chemistry*, **265**, 10181–4.

Henis, Y.I., Moustakas, A., Lin, H.Y. & Lodish, H.F. (1994). The type II and III TGF-β receptors form homo-oligomers. *Journal of Cell Biology*, **126**, 139–54.

Jennings, J.C., Mohan, S., Linkhart, T.A., Widstrom, R. & Baylink, D.J. (1988). Comparison of the biological actions of TGF beta-1 and TGF beta-2: differential activity in endothelial cells. *Journal of Cell Physiology*, **137**, 167–72.

Keller, J.R., Mantel, C., Sing, G.K., Ellingsworth, L.R., Ruscetti, S.K. & Ruscetti, F.W. (1988). Transforming growth factor beta 1 selectively regulates early murine hematopoietic progenitors and inhibits the growth of IL-3-dependent myeloid leukemia cell lines. *Journal of Experimental Medicine*, **168**, 737–50.

Kulkarni, A.B., Huh, C., Becker, D., Geiser, A., Lyght, M., Flanders, K.C., Roberts, A.B., Sporn, M.B., Ward, J.M. & Karlsson, S. (1993). Transforming growth factor-β1 null mutation in mice causes excessive inflammatory response and early death. *Proceedings of the National Academy of Sciences, USA*, **90**, 770–4.

Lehnert, S.A. & Akhurst, R.J. (1988). Embryonic expression pattern of TGF beta type-1 RNA suggests both paracrine and autocrine mechanisms of action. *Development*, **104**, 263–73.

Letterio, J.J., Gieser, A.G., Kulkarni, A.B., Roche, N.S., Sporn, M.B. & Roberts, A.B. (1994). Maternal rescue of the transforming growth factor β knockout. *Science*, **264**, 1936–8.

Lin, H.Y., Wang, D., Ng, E.E., Weinberg, R.A. & Lodish, H.F. (1992). Expression cloning of the TGF-beta type II receptor, a functional transmembrane serine/threonine kinase. *Cell*, **68**, 775–85.

Liu, J., Baker, J., Perkins, A.S., Robertson, E.J. & Efstradiadis, A. (1993). Mice carrying null mutations of the genes encoding insulin-like growth factor 1 (IGF-1) and type 1 IGF receptor (IGF1R). *Cell*, **75**, 59–72.

Madri, J.A., Bell, L. & Merwin, J.R. (1992). Modulation of vascular cell behaviour by TGF-β. *Molecular Reproductive Development*, **32**, 121–6.

Madri, J.A., Pratt, B.M. & Tucker, A.M. (1988). Phenotypic modulation of endothelial cells by transforming growth factor-β depends upon the composition and organization of the extracellular matric. *Journal of Cell Biology*, **106**, 1375–84.

Mansour, S.L., Thomas, K.R. & Capecchi, M.R. (1988). Disruption of the proto-oncogene int-2 in mouse embryo-derived stem cells: a general strategy for targeting mutations to non-selectable genes. *Nature*, **336**, 348–52.

Massague, J., Cheifetz, S., Endo, T. & Nadal-Ginard, B. (1986). Type β transforming growth factor is an inhibitor of myogenic differentiation. *Proceedings of the National Academy of Sciences, USA*, **83**, 8206–10.

Merwin, J.R., Newman, W., Beall, D., Tucker, A. & Madri, J.A. (1991). Vascular cells respond differentially to transforming growth factors-β1 and β2. *American Journal of Pathology*, **138**, 37–51.

Millan, F.A., Kondaiah, P., Denhez, F. & Akhurst, R.J. (1991). Embryonic gene expression patterns of TGF βs 1, 2 and 3 suggest different developmental functions *in vivo*. *Development*, **111**, 131–44.

Muller, G., Behrens, J., Nussbaumer, U., Bohlen, P. & Birchmeier, W. (1987). Inhibitory action of transforming growth factor β on endothelial cells. *Proceedings of the National Academy of Sciences, USA*, **84**, 5600–4.

Muslin, A.J. & Williams, L.T. (1991). Well-defined growth factors promote cardiac development in *axolotl* mesodermal explants. *Development*, **112**, 1095–101.

Ohta, M., Greenberger, J.S., Anklesaria, P., Bassols, A. & Massague, J. (1987). Two forms of transforming growth factor-β distinguished by multipotential haematopoietic progenitor cells. *Nature*, **329**, 539–41.

Olsen, E.N., Sternberg, E., Hu, J.S., Spizz, G. & Wilcox, C. (1986). Regulation of myogenic differentiation by type β transforming growth factor. *Journal of Cell Biology*, **103**, 1799–805.

Ottman, O. & Pelus, L. (1988). Differential proliferative effects of transforming growth factor β on human hematopoietic progenitor cells. *Journal of Immunology*, **140**, 2661.

Potts, J.D., Walder, J.A., Weeks, D.L. & Runyan, R.B. (1991). Epithelial–mesenchymal transformation of embryonic cardiac endothelial cells is inhibited by a modified antisense oligonucleotide to TGF-β3. *Proceedings of the National Academy of Sciences, USA*, **88**, 1516–20.

Qian, S.W., Burmester, J.K., Merwin, J.R., Madri, J.A., Sporn, M.B. & Roberts, A.B. (1992). Identification of a structural domain that distinguishes the actions of the type 1 and 2 isoforms of transforming growth factor β on endothelial cells. *Proceedings of the National Academy of Sciences, USA*, **89**, 6290–4.

Roberts, A.B., Anzano, M.A., Meyer, C.A., Wideman, J., Blacher, R., Pan, Y., Stein, S., Lehrman, S.R., Smith, J.M., Lamb, L.C. & Sporn, M.B. (1983). Purification and properties of a type β transforming growth factor from bovine kidney. *Biochemistry*, **22**, 5692–8.

Roberts, A.B. & Sporn, M.B. (1990). *Peptide Growth Factors and their Receptors – Handbook of Experimental Pharmacology*, ed. M.B. Sporn and A.B. Roberts, pp. 419–472. Heidelberg: Springer-Verlag.

Roberts, A.B., Vodovotz, Y., Roche, N.S., Sporn, M.B. & Nathan, C.F. (1992). Role of nitric oxide in antagonistic effects of transforming growth factor-β and interleukin-1β on the beating rate of cultured cardiac myocytes. *Molecular Endocrinology*, **6**, 1921–30.

Rosa, F., Roberts, A.B., Danielpour, D., Dart, L.L., Sporn, M.B. & Dawid, I.B. (1988). Mesoderm induction in amphibians: the role of TGF-β2-like factors. *Science*, **239**, 783–5.

Rosenquist, G.C. & de Haan, R.L. (1966). Migration of precardiac cells in the chick: A photoautoradiographic study. *Carnegie Institute Contributions to Embryology*, **38**, 111–21.

Shull, M.M., Ormsby, I., Kier, A.B., Pawlowski, S., Diebold, R.J., Yin, M., Allen, R., Sidman, C., Proetzel, G., Calvin, D., Annunziata, N. & Doetschman, T. (1992). Targeted disruption of the mouse transforming growth factor-β1 gene results in multifocal inflammatory disease. *Nature*, **359**, 693–9.

Sing, G.K., Keller, J.R., Ellingsworth, L.R. & Ruscetti, F.W. (1988). Transforming growth factor β selectively inhibits normal and leukemic human bone marrow cell growth *in vitro*. *Blood*, **72**, 1504–11.

Sing, G.K., Keller, J.R., Ellingsworth, L.R. & Ruscetti, F.W. (1989). Transforming growth factor-β1 enhances the suppression of human hematopoieses by tumor necrosis factor-alpha or recombinant interferon-alpha. *Journal of Cell Biochemistry*, **39**, 107–15.

Slager, H.G., van Inzen, W., Freund, E., van den Eijinden-van Raaij, A.J.M. & Mummery, C.L. (1993). Retinoic acid concentration-dependent differentiation of aggregated embryonic stem cells: effects of TGF-β isoforms on muscle formation. *Developments in Genetics*, **14**, 212–24.

Vaidya, T.B., Rhodes, S.J., Taparowsky, E.J. & Konieczny, S.F. (1989). Fibroblast growth factor and transforming growth factor-β

repress transcription of the myogenic regulatory gene MyoD1. *Molecular and Cellular Biology*, **9**, 3576–9.

Wilkinson, D.G., Bailes, J.A., Champion, J.E. & McMahon, A.P. (1987). A molecular analysis of mouse development from 8 to 10 days *post coitum* detects changes only in globin expression. *Development*, **99**, 493–500.

Wrana, J.L., Attisano, L., Weiser, R., Ventura, F. & Massague, J. (1994). Mechanism of activation of the TGF-β receptor. *Nature*, **370**, 341–7.

Yamaguchi, T.P., Dumont, D.J., Conlon, R.A., Breitman, M.L. & Rossant, J. (1993). *flk-1*, an *flt*-related receptor tyrosine kinase is an early marker for endothelial cell precursors. *Development*, **118**, 489–98.

Zentella, A. & Massague, J. (1992). Transforming growth factor β induces myoblast differentiation in the presence of mitogens. *Proceedings of the National Academy of Sciences, USA*, **89**, 5176–80.

ELEANOR J. MACKIE and SUSAN RAMSEY

Tenascin: an extracellular matrix protein associated with bone growth

Introduction

Bone growth, whether longitudinal or appositional, requires a source of differentiated osteoblasts capable of secreting the proteins of bone matrix. Thus, factors influencing both the proliferation and differentiation of osteoblast precursors are important for bone growth. Cells of the osteoblast lineage respond to many environmental influences including soluble factors (hormones, growth factors) and mechanical loading. The proteins of the extracellular matrix, as well as having important structural roles in bone, may also be important local regulators of bone cell function, and may act as long-term insoluble mediators of the actions of soluble factors or load.

In recent years, it has been demonstrated that extracellular matrix proteins, acting through cell-surface receptors such as integrins, can have potent specific effects on cell behaviour, being able to influence cell proliferation, migration and differentiation. The extracellular matrix protein tenascin-C is of particular interest in bone development and growth because of its selective association with the developing skeleton. Tenascin-C (hereafter referred to simply as 'tenascin') is a member of the tenascin gene family, composed of at least four members (Chiquet-Ehrismann, Hagios & Matsumoto, 1994). Tenascin is a large hexameric glycoprotein with disulphide-linked subunits, each consisting of several structural domains, including epidermal growth factor-like repeats, fibronectin type III repeats and a fibrinogen-like terminal domain (Fig. 1; Spring, Beck & Chiquet-Ehrismann, 1989). Alternative splicing of the tenascin gene results in the existence of differently sized subunits, the number of possible subunits varying between species. The larger splice variants differ from the smaller variants by the presence of additional fibronectin type III repeats. Different subunits exhibit cell- and tissue-specific expression patterns (Matsuoka *et al.*, 1990; Prieto *et al.*, 1990; Mackie & Tucker, 1992). The multidomain structure

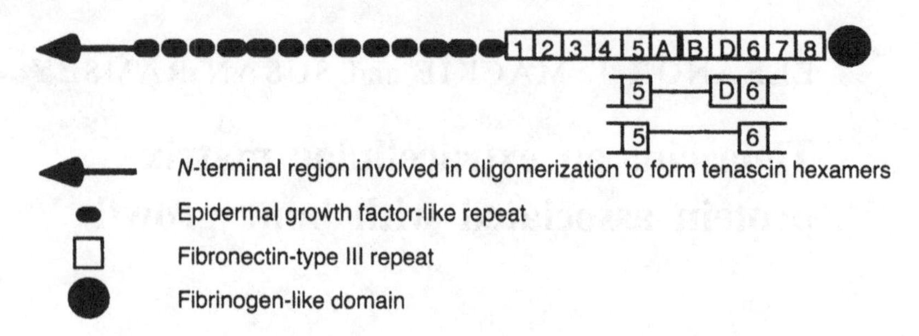

Fig. 1. Schematic representation of the structure of the three major subunits of chicken tenascin found in developing skeletal tissues. The smaller two splice variants result from exclusion of fibronectin type III repeats A and B, or A,B and D.

of tenascin and the existence of different splice variants suggest that this protein may have multiple sites of interaction with cells and other matrix proteins, and is likely to have different functions in different tissues. Such suggestions are supported by recent studies aimed at identifying functional regions of the tenascin molecule (Spring *et al.*, 1989; Joshi *et al.*, 1993; Sriramarao, Mendler & Bourdon, 1993).

Tenascin is selectively associated with tissue remodelling in many embryonic and pathological tissues. The importance of tenascin in central nervous system development is illustrated by its ability to promote neurite outgrowth and cerebellar granule cell migration (Husmann, Faissner & Schachner, 1992). In the connective tissues, few functions for tenascin have so far been described. Tenascin stimulates chondrogenesis in limb bud mesenchymal cell cultures (Mackie, Thesleff & Chiquet-Ehrismann, 1987). Tenascin also appears to be important for skin morphogenesis, since anti-tenascin inhibits chicken feather bud elongation in organ culture (Jiang & Choung, 1992). In addition, either stimulatory or inhibitory effects of tenascin on cell proliferation have been reported in a variety of cell types (Chiquet-Ehrismann *et al.*, 1986; Crossin, 1991; End *et al.*, 1992).

Many different cell types adhere to tenascin *in vitro*. Contrary to many other adhesive macromolecules, tenascin does not confer spreading but allows adherent cells to remain round (Chiquet-Ehrismann *et al.*, 1988; Mackie *et al.*, 1992). Thus the strength of adhesion of cells to tenascin is relatively weak (Lotz *et al.*, 1989). The effect of tenascin on cell shape may be critical for some of its functions. Chondrocyte differentiation, for example, can be stimulated by culturing mesenchymal cells under conditions which cause cell rounding, and is reversed by forcing cells to spread on their substratum (Benya & Shaffer, 1982).

Adhesion of cells to tenascin although weak, is specific. Two different integrins ($\alpha_v\beta_3$ and $\alpha_2\beta_1$) have been identified as tenascin receptors in some cells (Salmivirta *et al.*, 1991; Joshi *et al.*, 1993; Sriramarao *et al.*, 1993). There is also strong evidence that cell-surface heparan sulphate proteoglycans can function as tenascin receptors in connective tissue cells (Aukhil *et al.*, 1993). Different receptors are recognized by different regions of tenascin and are likely to mediate distinct functions.

Expression of tenascin by bone cells

During bone development, tenascin is first expressed by cells of the mesenchymal lineage as they start to undergo chondrogenic or osteoblastic differentiation (Mackie *et al.*, 1987). With progressive chondrocyte differentiation, tenascin expression is lost, and thus tenascin is absent from mature cartilage matrix. At sites of intramembranous ossification, tenascin is detectable by immunohistochemistry in association with cells of the osteoblast lineage. In endochondral bones, although absent from hypertrophic cartilage, strong tenascin staining is seen in association with the invading osteogenic cells forming the primary centre of ossification (Mackie *et al.*, 1987). Tenascin staining appears slightly earlier in preosteoblastic cells of the periosteum than does alkaline phosphatase activity, a marker of the osteoblast phenotype (Väkevä *et al.*, 1990).

In growing post-natal bones, tenascin is absent from mineralized bone matrix. It continues to be detectable, however, on all bone surfaces. Fig. 2 shows immunoperoxidase staining of a cryosection of sub-metaphyseal cortical bone from the tibia of a ten-week-old rat. This bone is in the process of being remodelled from the primary spongiosa of the growing metaphysis (resulting from replacement of growth plate cartilage by bone) into the cortical bone of the diaphysis. The periosteum contains many osteoblast precursors actively undergoing osteoblast differentiation. Anti-tenascin staining of these cells is strongly positive. Tenascin is also seen on the surfaces of the recently formed Haversian canals and lining osteocyte lacunae. Tenascin is, however, clearly absent from the mineralized bone matrix. The section shown in Fig. 2 was taken from bone that had been demineralized prior to embedding, but similar results are obtained with sections of non-demineralized bone, indicating that the failure to detect tenascin in mineralized bone matrix is not an artefact resulting from demineralization.

The observation that tenascin is associated with cells of the osteoblast lineage in developing bone led to the investigation of tenascin

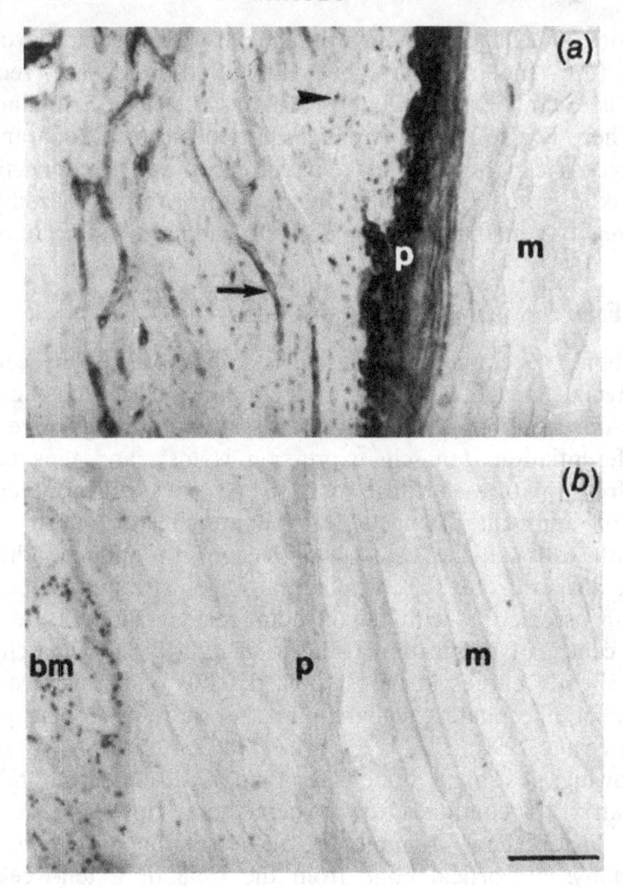

Fig. 2. The distribution of tenascin in bone undergoing rapid turnover. (*a*) Cryosection of demineralized sub-metaphyseal cortical bone from a 10-week-old rat, stained with rabbit anti-rat tenascin followed by the peroxidase–antiperoxidase detection system. (*b*) Cryosection adjacent to that shown in (*a*), stained with normal rabbit serum instead of anti-tenascin, as negative control. The bone marrow which is present in this field, but not in the field shown in (*a*), shows strong non-specific staining. m, muscle; p, periosteum; bm, bone marrow, arrow, a Haversian canal, arrowhead, osteocyte. Bar = 200 μm.

expression by osteoblastic cells in culture. The ROS 17/2.8 cell line is derived from a rat osteosarcoma and has been defined as osteoblast-like because it expresses a range of characteristics typical of the osteoblast phenotype, including elevated alkaline phosphatase activity and responsiveness to parathyroid hormone (Rodan & Rodan, 1984). ROS 17/2.8 cells have been shown to synthesize and secrete tenascin into their medium (Mackie & Tucker, 1992). The tenascin secreted by ROS 17/2.8 cells contains predominantly the largest splice variant. Similarly, primary osteoblast-enriched cultures from chicken embryo calvarial bones secrete almost exclusively the largest splice variant. This is of interest since the periosteal fibroblast-like cells from which the osteoblasts are derived secrete all of the three major chicken tenascin splice variants (Mackie & Tucker, 1992). Results of *in situ* hybridization studies with a cDNA probe specific for the large splice variant indicate that this pattern of cell-specific expression of variants also occurs *in vivo* (Mackie & Tucker, 1992). In contrast, cells undergoing chondrogenic differentiation, derived from the same mesenchymal cell population as osteoblasts, restrict their tenascin expression to the smaller splice variants (Vaughan *et al.*, 1987; Mackie & Tucker, 1992). The restricted expression by osteoblasts of the large tenascin splice variant suggests that the additional fibronectin type III repeats in this variant (see Fig. 1) confer some function related specifically to osteoblast behaviour.

Once cultured osteoblast-like cells reach confluence they start to deposit tenascin in their extracellular matrix. Immunofluorescence staining of overconfluent ROS 17/2.8 cells shows a fibrillar pattern of staining on the cell surface (Fig. 3). Primary chicken osteoblast-enriched cultures showed a similar fibrillar deposition of tenascin, specifically in regions of the culture that stained positively for endogenous alkaline phosphatase activity (Mackie & Tucker, 1992). Thus osteoblasts express tenascin and deposit it in their extracellular matrix. This indicates, as would be expected from tissue distribution studies, that cells of the osteoblast lineage are a major source of the tenascin found in bone tissue. As mentioned above, *in situ* hybridization studies have confirmed that tenascin is expressed by osteoblastic cells in embryonic chicken long bones (Mackie & Tucker, 1992).

Stimulation of osteoblast differentiation by tenascin

Since tenascin is generally absent from mineralized bone matrix, it cannot possibly play a structural role in bone. The fact that tenascin is secreted by cells of the osteoblast lineage and remains in their immediate environment *in vivo* strongly suggests that tenascin is important for the regulation of osteoblast behaviour. For this reason, the ability of tenascin to stimulate osteoblast differentiation has been investigated.

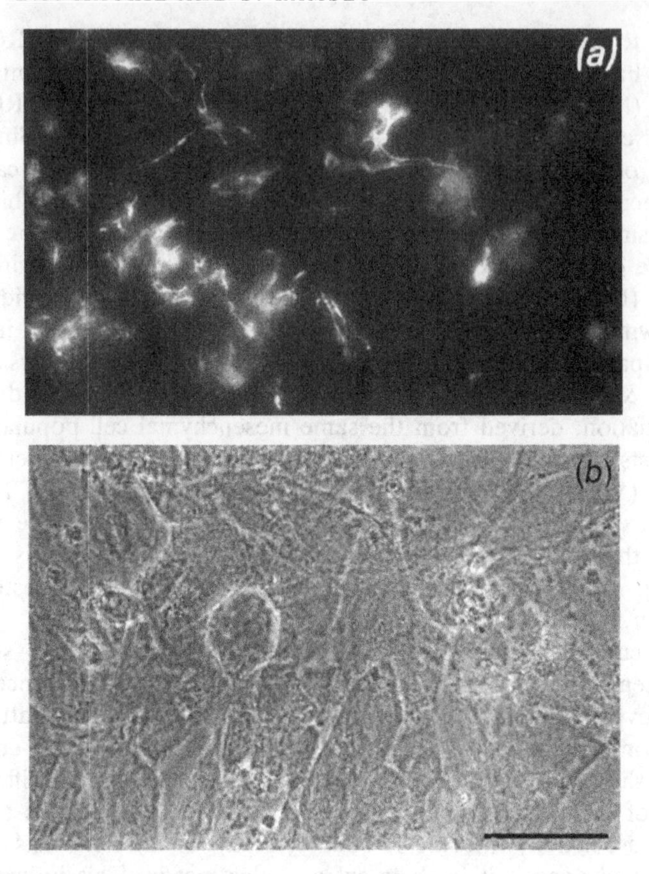

Fig. 3. Deposition of tenascin in the matrix of osteoblast-like cells in culture. (*a*) Overconfluent ROS 17/2.8 cells stained by indirect immunofluorescence with anti-rat tenascin. (*b*) The same field as that shown in (*a*), visualized by phase contrast. Bar = 50 μm.

Initial experiments were aimed at investigating whether tenascin interacts specifically with osteoblasts. The ability of ROS 17/2.8 cells to adhere to tenascin was quantitated in adhesion assays. Wells of 96-well immunoassay plates were coated overnight with tenascin or bovine serum albumin (as negative control). Adhesion of cells to the coated substrata over a two-hour period was quantitated as toluidine blue uptake into adherent cells, measured spectrophotometrically. As shown in Fig. 4, ROS 17/2.8 cells adhered specifically to tenascin, as compared with the BSA control. The morphology of cells plated on tenascin was also investigated and compared with that of cells plated

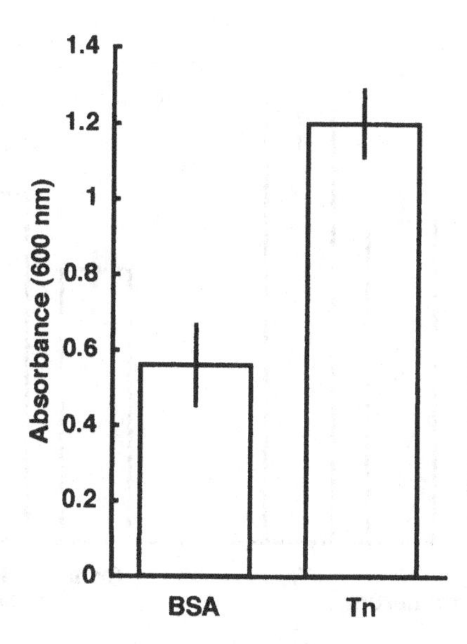

Fig. 4. Adhesion of ROS 17/2.8 cells to tenascin (Tn) or bovine serum albumin (BSA) as negative control substratum. Results are expressed as absorbance due to toluidine blue uptake in adherent cells (mean $\pm$ standard error, $n = 4$).

on fibronectin, another extracellular matrix glycoprotein. Fibronectin is present in bone, but has a much more widespread distribution in the connective tissues than does tenascin, and thus seems unlikely to have a function specific for bone tissue. Cells that adhered to tenascin remained rounded, as compared to cells plated on fibronectin, which spread and flattened. Thus the effect of tenascin on osteoblast morphology is similar to that seen in other cell types.

Endogenous alkaline phosphatase activity is a marker of the osteoblast phenotype. Factors stimulating alkaline phosphatase activity are regarded as being able to stimulate osteoblast differentiation. The ability of a tenascin-coated substratum to stimulate alkaline phosphatase activity in ROS 17/2.8 cells was therefore investigated. Tissue culture substrata were coated with tenascin or fibronectin or left uncoated, as negative control. Cells were plated on these substrata in the presence or absence of fetal calf serum and incubated for 24 hours; alkaline phosphatase activity in the cell layer was then determined and corrected for the protein content of the cell layer. Cells plated on tenascin in the absence of serum showed two-fold higher alkaline phosphatase activity than control cells (Fig. 5). Alkaline phosphatase activity was

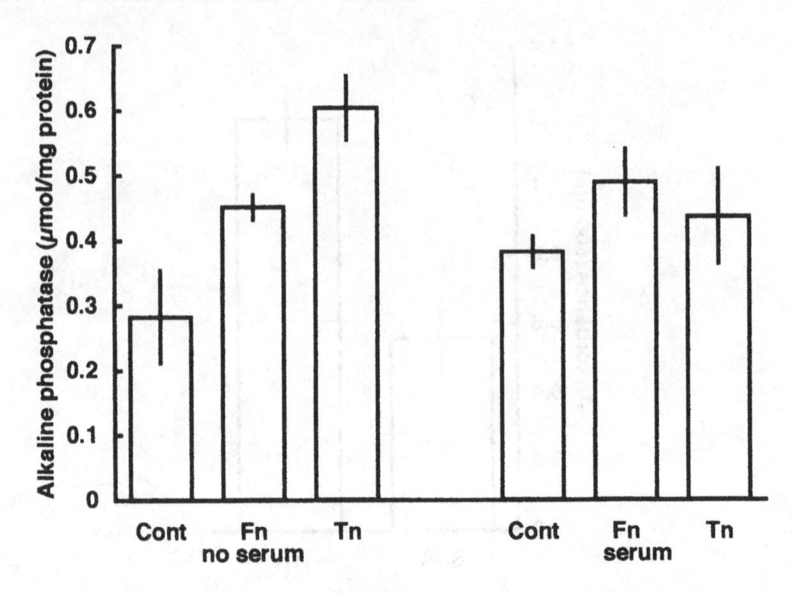

Fig. 5. Alkaline phosphatase activity in ROS 17/2.8 cells adherent to tenascin (Tn), fibronectin (Fn) or uncoated plastic (Cont), cultured in the absence or presence of fetal calf serum (10% v/v). Results are expressed as total alkaline phosphatase activity of the cell layer, corrected for protein content of the cell layer (mean ± standard error, $n = 4$).

also slightly elevated in cells plated on fibronectin. Cells cultured in the presence of serum showed no significant change in their alkaline phosphatase activity in response to either of the extracellular matrix substrata.

These results clearly demonstrate that tenascin is able to stimulate expression of one marker of the osteoblast phenotype. The inability of tenascin to influence alkaline phosphatase activity in the presence of fetal calf serum suggests that serum contains factors able to override the stimulatory effect of tenascin. Serum stimulates cell proliferation, and the degree of osteoblast differentiation is inversely correlated with proliferation. It seems likely that growth factors present in serum, by inducing proliferation, prevent ROS 17/2.8 cells from responding to differentiation-inducing stimuli.

The role of tenascin in osteoblast function

The stimulation by substratum-attached tenascin of alkaline phosphatase activity in osteoblast-like cells indicated that tenascin is able to stimulate osteoblast differentiation. These results, together with results of the tissue distribution studies of tenascin in growing bones, suggest that tenascin is an important regulator of osteoblast differentiation *in vivo*. Tenascin is abundant in the local environment of periosteal cells undergoing osteoblast differentiation. This association is seen throughout skeletogenesis, beginning at the onset of intramembranous bone formation. During post-natal growth, bones are continually modelled by osteoclastic resorption and osteoblastic formation to maintain the shape of the bone as it elongates and grows in circumference. In sites of rapid turnover due to modelling (such as the sub-metaphyseal bone shown in Fig. 2) the association of strong tenascin expression with the process of osteoblast differentiation is maintained.

Many soluble factors, produced locally or systemically, are known to be able to influence osteoblast function. Examples include hormones (such as parathyroid hormone and 1,25-dihydroxyvitamin D_3), and growth factors (including the insulin-like growth factors I and II; Puzas, 1993). Another regulator of bone cell activity, extremely important in the development and maintenance of appropriate bone architecture, is mechanical load. Tenascin, as an extracellular matrix protein presented to cells in insoluble form, is likely to provide local signals to differentiating osteoblasts over a longer term than is possible for soluble factors or load. It seems likely that tenascin may help to mediate some of the responses of osteoblasts to initiating factors that themselves have much shorter half-lives than tenascin. It will be of interest to investigate which soluble factors are able to regulate tenascin expression in cultured osteoblasts.

There is considerable evidence, based on tissue distribution studies in several tissues, that tenascin expression is stimulated by mechanical strain. Tenascin is strongly expressed at the interfaces between elements of the musculoskeletal system, and in the vascular wall at sites of branching (Thesleff *et al.*, 1988; Mackie *et al.*, 1992). The fact that a single period of dynamic loading stimulates measurable new bone formation has led to the concept of 'strain memory' (Lanyon, 1992). This refers to the idea that mechanical loading induces changes in bone cells or their environment that allow them to continue to respond to a stimulus after it is no longer active. An extracellular matrix protein such as tenascin, retained over days (instead of minutes or hours) in the osteoblast environment, seems ideally suited to playing this role.

The mechanism of tenascin's stimulation of osteoblastic alkaline phosphatase activity has not yet been investigated. It is likely that tenascin exerts its effects through a receptor of the integrin family or a cell surface heparan sulphate proteoglycan, since these are the two types of receptor known to be involved in adhesion of connective tissue cells to tenascin (Aukhil *et al.*, 1993; Joshi *et al.*, 1993; Sriramarao *et al.*, 1993). One possibility is that tenascin influences osteoblast differentiation through its effect on cell shape, rather than through a direct, receptor-mediated regulation of gene expression. Differentiated osteoblasts are characteristically rounded cells; it is possible that cell rounding alone is sufficient to induce differentiation, as is the case for chondrocytes (Benya & Shaffer, 1982).

Tenascin is expressed in growing bones in association with the process of osteoblast differentiation. *In vitro*, tenascin is able to interact specifically with osteoblasts, altering their morphology and stimulating osteoblast differentiation. In conclusion, tenascin appears to be an important local regulator of osteoblast activity in growing bones *in vivo*.

References

Aukhil, I., Joshi, P., Yan, Y. & Erickson, H.P. (1993). Cell- and heparin-binding domains of the hexabrachion arm identified by tenascin expression proteins. *Journal of Biological Chemistry*, **268**, 2542–53.

Benya, P.D. & Shaffer, J.D. (1982). Dedifferentiated chondrocytes reexpress the differentiated collagen phenotype when cultured in agarose gels. *Cell*, **30**, 215–24.

Chiquet, Ehrismann, R., Hagios, C. & Matsumoto, K. (1994). The tenascin gene family. *Perspectives on Developmental Neurobiology*, **2**, 3–7.

Chiquet-Ehrismann, R., Kalla, P., Pearson, C.A., Beck, K. & Chiquet, M. (1988). Tenascin interferes with fibronectin action. *Cell*, **53**, 383–90.

Chiquet-Ehrismann, R., Mackie, E.J., Pearson, C.A. and Sakakura, T. (1986). Tenascin: an extracellular matrix protein involved in tissue interactions during fetal development and oncogenesis. *Cell*, **47**, 131–9.

Crossin, K.L. (1991). Cytotactin binding: inhibition of stimulated proliferation and intracellular alkalinization in fibroblasts. *Proceedings of the National Academy of Sciences, USA*, **88**, 11403–7.

End, P., Panayotou, G., Entwistle, A., Waterfield, M.D. & Chiquet, M. (1992). Tenascin: a modulator of cell growth. *European Journal of Biochemistry*, **209**, 1041–51.

Husmann, K., Faissner, A. & Schachner, M. (1992). Tenascin promotes cerebellar granule cell migration and neurite outgrowth by different domains in the fibronectin type III repeats. *Journal of Cell Biology*, **116**, 1475–86.

Jiang, T.-X. & Chuong, C.-M. (1992). Mechanism of skin morphogenesis. I. Analyses with antibodies to adhesion molecules tenascin, N-CAM, and integrin. *Developmental Biology*, **150**, 82–98.

Joshi, P., Chung, C.-Y., Aukhil, I. & Erickson, H.P. (1993). Endothelial cells adhere to the RGD domain and the fibrinogen-like terminal knob of tenascin. *Journal of Cell Science*, **106**, 389–400.

Lanyon, L.E. (1992). Control of bone architecture by functional load bearing. *Journal of Bone and Mineral Research*, **7**, S369–75.

Lotz, M.M., Burdsal, C.A., Erickson, H.P. & McClay, D.R. (1989). Cell adhesion of fibronectin and tenascin: quantitative measurements of initial binding and subsequent strengthening response. *Journal of Cell Biology*, **109**, 1795–805.

Mackie, E.J., Scott-Burden, T., Hahn, A.W.A., Kern, F., Bernhardt, J., Regenass, S., Weller, A. & Bühler, F.R. (1992). Expression of tenascin by vascular smooth muscle cells: alterations in hypertensive rats and stimulation by angiotensin II. *American Journal of Pathology*, **141**, 377–88.

Mackie, E.J., Thesleff, I. & Chiquet-Ehrismann, R. (1987) Tenascin is associated with chondrogenic and osteogenic differentiation *in vivo* and promotes chondrogenesis *in vitro*. *Journal of Cell Biology*, **105**, 2569–79.

Mackie, E.J. & Tucker, R.P. (1992). Tenascin in bone morphogenesis: expression by osteoblasts and cell type-specific expression of splice variants. *Journal of Cell Science*, **103**, 765–71.

Matsuoka, Y., Spring, J., Ballmer-Hofer, K., Hofer, U. & Chiquet-Ehrismann, R. (1990). Differential expression of tenascin splicing variants in the chick gizzard and in cell cultures. *Cell Differentiation and Development*, **32**, 417–24.

Prieto, A.L., Jones, F.S., Cunningham, B.A., Crossin, K.L. & Edelman, G.M. (1990). Localization during development of alternatively spliced forms of cytotactin mRNA by *in situ* hybridization. *Journal of Cell Biology*, **111**, 685–98.

Puzas, J.E. (1993). The osteoblast. In *Primer on the Metabolic Bone Diseases and Disorders of Mineral Metabolism*, ed. M.J. Favus, pp. 15–21. New York: Raven Press.

Rodan, S.B. & Rodan, S.B. (1984). Expression of the osteoblastic phenotype. In *Advances in Bone and Mineral Research Annual II*, ed. W.A. Peck, pp. 244–285. Amsterdam: Excerpta Medica.

Salmivirta, M., Elenius, K., Vainio, S., Hofer, U., Chiquet-Ehrismann, R., Thesleff, I. & Jalkanen, M. (1991). Syndecan from embryonic tooth mesenchyme binds tenascin. *Journal of Biological Chemistry*, **266**, 7733–9.

Spring, J., Beck, K. & Chiquet-Ehrismann, R. (1989). Two contrary functions of tenascin: dissection of the active sites by recombinant tenascin fragments. *Cell*, **59**, 523–34.

Sriramarao, P., Mendler, M. & Bourdon, M.A. (1993). Endothelial cell attachment and spreading on human tenascin is mediated by $\alpha_2\beta_1$ and $\alpha_v\beta_3$ integrins. *Journal of Cell Science*, **105**, 1001–12.

Thesleff, I., Kantomaa, T., Mackie, E.J. & Chiquet-Ehrismann, R. (1988). Immunohistochemical localization of the matrix glycoprotein tenascin in the skull of the growing rat. *Archives of Oral Biology*, **33**, 383–90.

Väkevä, L., Mackie, E.J., Kantomaa, T. & Thesleff, I. (1990). Comparison of the distribution patterns of tenascin and alkaline phosphatase in developing teeth, cartilage, and bone of rats and mice. *Anatomical Record*, **228**, 69–76.

Vaughan, L., Huber, S., Chiquet, M. & Winterhalter, K.H. (1987). A major, six-armed glycoprotein from embryonic cartilage. *EMBO Journal*, **6**, 349–53.

JOHN HESKETH

Compartmentation of protein synthesis, mRNA targeting and c-myc expression during muscle hypertrophy and growth

Introduction

Post-natal muscle growth, which is one of the major growth processes of the mammalian body, involves a combination of myofibre elongation and an increase in fibre diameter. The latter occurs by hypertrophy (increase in size without cell division) whilst muscle elongation is achieved by addition of extra cell units. Muscle mass also increases in response to workload, to stretch and to anabolic agents, and in all cases the predominant myotrophic response involves hypertrophy. This requires increased synthesis of myofibrillar protein and a rapid growth and assembly of the contractile myofibrils whilst structural integrity and physiological function of the sarcomeres is maintained. During muscle hypertrophy the increased amounts of newly synthesized protein must be targeted to specific parts of the fibre to ensure that a spatially precise myofibrillar assembly can occur.

All cells target proteins to specific intracellular sites but the organization of the synthesis of the myofibrillar proteins, particularly large proteins such as titin and myosin, presents the long and ordered myofibre with unique logistic problems. These problems of cell organization occur during normal turnover of the myofibrillar proteins and myofibril repair, both of which require the newly synthesized proteins to be incorporated into the contractile apparatus, but are exacerbated during hypertrophy. Thus the intracellular targeting of newly synthesized protein is crucial to ordered muscle hypertrophy and to maintenance of the myofibrillar structure. As illustrated in Fig. 1, this targeting could theoretically be achieved by targeting of the synthesized protein, by spatial organization of ribosomes and mRNAs so that synthesis occurs close to the protein's site of function or by co-translational assembly. In addition, the myofibre is multi-nucleated and control of the large cytoplasmic volume is maintained by hundreds of myonuclei along its length. There is evidence for variation in gene expression

(*a*) Protein targeting

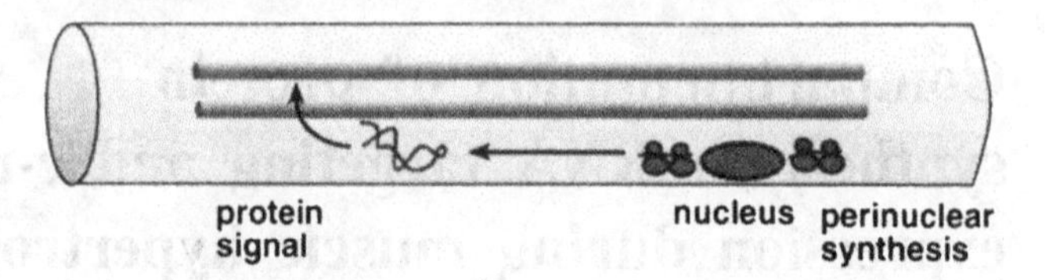

(*b*) mRNA targeting

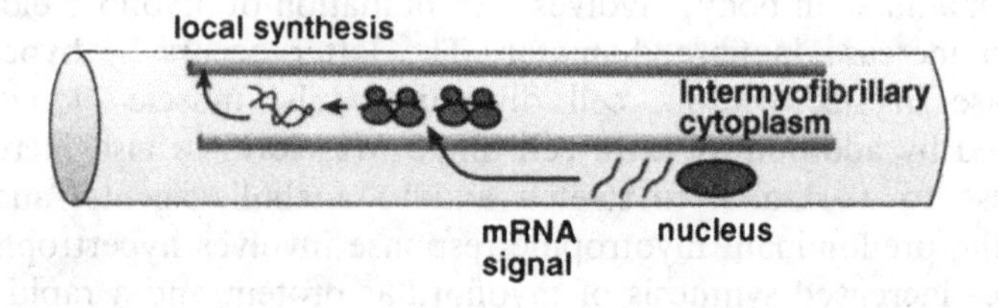

(*c*) Co-translational

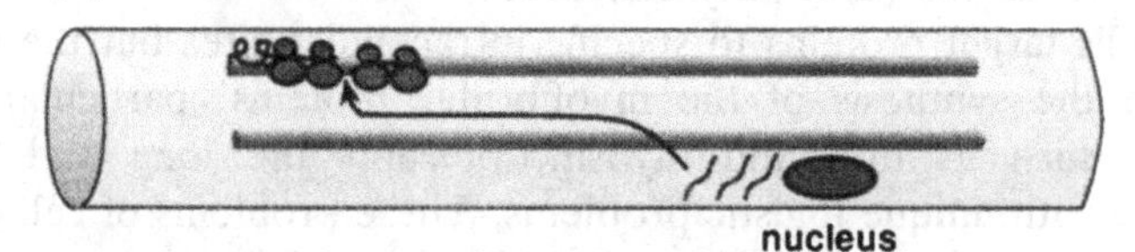

Fig. 1. Possible mechanisms for targeting of myofibrillar proteins.

within the individual myonuclei (Hall & Ralston, 1989), and this raises the possibility that differential nuclear activation may contribute to the spatial control of synthetic activity during hypertrophy.

This chapter addresses the problem of how coordinated growth is achieved during myofibre hypertrophy, particularly how myofibrillar proteins are targeted to their site of function, and whether it involves spatial organization of the protein synthetic apparatus.

Ribosome and mRNA distribution in skeletal muscle

Electron microscopy of skeletal muscle has shown consistently that ribosomes are present in the subsarcolemmal cytoplasm, especially the paranuclear cones (Galavazi, 1971; Gauthier & Schaeffer, 1974). In addition, several reports have demonstrated that there are ribosomes in close association with the myofibrils (Larsen, Hudgson & Walton, 1969; Galavazi, 1971; Gauthier & Mason-Savas, 1993). Immunohistochemical studies with antibodies raised against ribosomal 60S subunits have also revealed that, in addition to strong staining of the subsarcolemmal cytoplasm, there is also staining of the myofibrillar region of the fibres (Horne & Hesketh, 1990*a*), a result consistent with ribosomes being present in both the subsarcolemmal and myofibrillary cytoplasm.

Early electron micrographs showed ribosomes present in the intermyofibrillary cytoplasm and in association with either the I-band (Galavazi, 1971) or the A-band (Larsen *et al.*, 1969). Image analysis of skeletal muscle sections labelled with anti-ribosome antibodies suggested a regular association of ribosomes with the myofibrils with the labelling corresponding to the myosin-containing A-bands (Home & Hesketh, 1990*a*). Recent electron micrographs of chicken muscle show ribosomes to be present both in the *intra*myofibrillary cytoplasm of the A-bands and outside the myofibril, but associated with the I-band rather than the A-band (Gauthier & Mason-Savas, 1993). In cardiac muscle, immunohistochemical staining with the same anti-60S ribosomal antibody discussed above again showed a regular association of ribosomal material with the myofibrils but, in this case, the ribosomes appeared to co-localize with desmin and α-actinin, consistent with their concentration close to the Z-disc within the I-band (Larsen *et al.*, 1994). Ribosome localization at intercalated discs and the Z-disc was also demonstrated by electron microscopy and immunogold labelling. Although these studies all suggest that ribosomes are present in the intermyofibrillar cytoplasm and are associated with the myofibrils, there is some variation in the precise location observed.

Although some of the variation in ribosome distribution may be due to experimental artefact, some may have a physiological basis; either from differences in ribosome distribution between muscle fibres of different metabolic characteristics, or from variation in the physiological state, and thus the pattern of protein synthesis, of the muscles under study. The former may be especially relevant when comparing data from cardiac and skeletal muscle since the skeletal muscles studied (psoas, plantaris) were predominantly composed of 'fast' myofibres, whilst cardiac muscle fibres have a more oxidative metabolism and slower contractile characteristics. Alternatively, the fluctuating physiological requirements of muscles will dictate alterations in the patterns of proteins being synthesized, and this in turn may lead to changes in the distribution of ribosomes. Indeed, there is evidence that ribosome distribution changes under conditions of altered synthetic rates. For example, less ribosomal material is found associated with the myofibrils in muscles from rats of 51 days of age compared to those of 14 days (Horne & Hesketh, 1990*a*) and this change occurs over a period when there is a considerable decrease in both the overall protein synthetic rate and the relative proportion of myofibrillar to sarcoplasmic protein synthesis (Waterlow, Garlick & Millward, 1978; Lewis, Kelly & Goldspink, 1984). Furthermore, the induction of the fast myosin isoform by curare was associated with a 2-fold increase in the number of ribosomes found in rows between the thick filaments in the A-band (Gauthier & Mason-Savas, 1993).

Thus, at periods of higher actomyosin or myosin synthesis, a greater proportion or ribosomes appear associated with the myofibrils. This suggests that the ribosomes in the intermyofibrillary cytoplasm synthesize myofibrillar proteins such a myosin locally, close to the point of insertion into the myofibrils. The hypothesis of local synthesis is compatible both with the presence of polysomes in or around the A-band (Larsen *et al.*, 1969) and the observation that during translation nascent myosin heavy chains in polysomes are associated with filaments of the cell matrix (Isaacs & Fulton, 1987) and are released by cytochalasin D, which depolymerizes the actin-containing microfilaments.

If myofibrillar proteins, such as myosin, are synthesized close to their site of assembly, one would expect to find localized distribution of mRNAs for some or all of the myofibrillar proteins. Indeed, there is evidence for localization of specific mRNAs in different cytoplasmic domains of myoblasts grown in culture (Lawrence & Singer, 1986; Hill & Gunning, 1993). In addition, hybridization studies *in situ* show that the myosin heavy chain mRNA is present, albeit at a low concentration, in the myofibrillar cytoplasm (Hesketh, Campbell &

Fig. 2. Myosin heavy chain mRNA distribution revealed by *in situ* hybridization. Myosin heavy chain slow isoform mRNA was detected in both longitudinal (L) and transverse (T) sections through rat soleus muscle using ^{35}S-labelled riboprobes. The autoradiographic grain density was quantified by microdensitometry (Hesketh *et al.*, 1991*a*) and the distribution calculated as shown in the Figure. m, myofibrillar cytoplasm; s, subsarcolemmal cytoplasm. Bar = 32 μm.

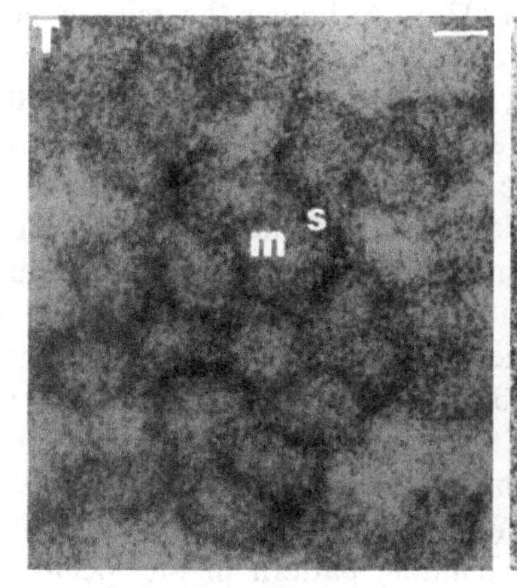

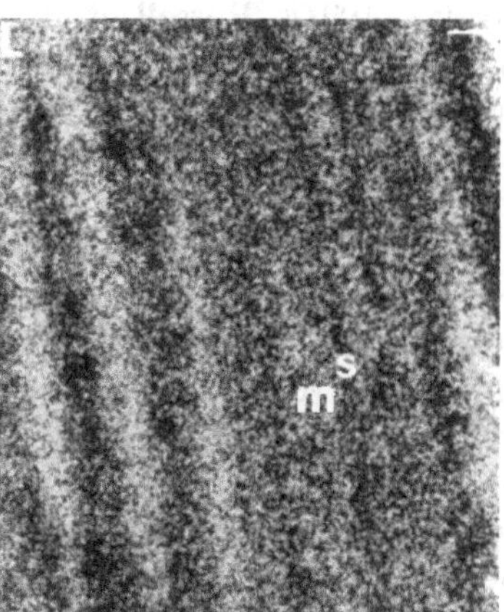

	Grain density	
	Transverse sections	Longtitudinal sections
Subsarcolemmal region (s)	6.0±0.4	5.6±0.3
Myofibrillar region (m)	3.9±0.4	2.5±0.3
Myofibrillar/subsarcolemmal ratio	0.65	0.45

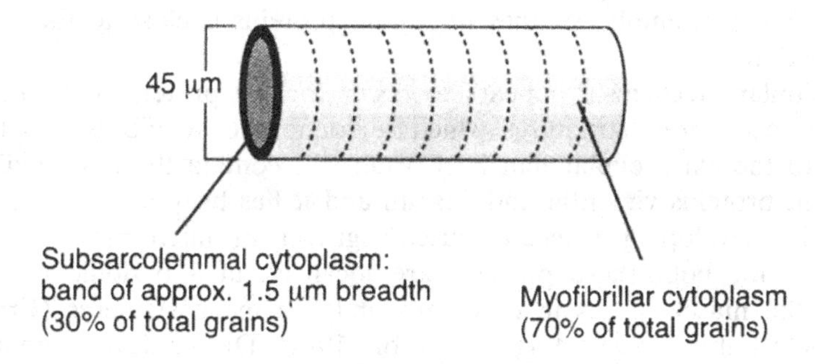

Loveridge, 1991*a*; Russell *et al.*, 1992*a*). Although the myosin heavy chain mRNA is present at a higher concentration in the subsarcolemmal cytoplasm (Aigner & Pette, 1990; Pomeroy *et al.*, 1991; Hesketh *et al.*, 1991*a*), quantification of the autoradiographic grain densities by microdensitometry showed that the majority of the mRNA was in the myofibrillar cytoplasm (Fig. 2; Hesketh *et al.*, 1991*a*). Assuming muscle fibres are cylinders of average diameter 45 μm with a subsarcolemmal cytoplasmic rim of 1.5 μm width, then by simple geometry the proportions of myosin mRNA in the myofibrillar and subsarcolemmal cytoplasm were calculated as 70 : 30. The one report of *in situ* hybridization showing clear localization of myosin heavy chain mRNA in the I-band is from a study of transforming fibres (Aigner & Pette, 1990). Although other reports show occasional vague banding, the bulk of experimental data shows no clear banding pattern in the localization nor that the mRNA is present within the myofibrils (Dix & Eisenberg, 1988; Hesketh *et al.*, 1991*a*; Pomeroy *et al.*, 1991; Russell, Wenderoth & Goldspink, 1992*b*).

It appears therefore that myosin heavy chain mRNA is found in the intermyofibrillary cytoplasm between the myofibrils but not within the myofibrils. There appears to be no preferential association of the myosin mRNA with the A-bands (myosin) of the myofibrils and so it seems unlikely that translation of the message is linked directly to assembly. However, the available data indicate that the majority of both ribosomes and myosin heavy chain mRNA are present in the intermyofibrillary cytoplasm, suggesting that it is in this region that myosin (and presumably other myofibrillar proteins) are synthesized. This implies that there is some targeting or spatial organization of the protein synthetic apparatus such that, although there is no co-translational assembly, synthesis of these proteins is close to their site of function.

A similar mechanism appears to occur for the protein components of the costameres, structures which surround the myofibrils and link them to the extracellular matrix. Costameres contain the intermediate filament proteins vimentin and desmin and it has been shown recently that, in developing chicken muscles grown in micromass culture, mRNAs for both these proteins are localized in a periodic fashion along the muscle fibres in close proximity to the costameres (Cripe, Morris & Fulton, 1993; Morris & Fulton, 1994). During development of the muscle the distribution of both mRNAs changed as the costameres appeared, and at the various stages of development they were distributed close to the site of protein localization. This appears therefore, to be a further example of how muscle structural proteins are synthesized locally and in a spatially organized way. In this case it has been demonstrated that mRNA localization is controlled in a precise manner during costamere formation and this highlights the potential importance of mRNA and protein targeting mechanisms in ensuring a controlled structural integrity during the growth, development and repair of muscle.

Ribosome and mRNA distribution in hypertrophying muscle

Rates of myofibrillar protein synthesis are particularly high during periods of rapid hypertrophy, for example that induced by administration of anabolic agents, by the compensatory workload and stretching that occurs following severance of the tendon to a synergistic muscle (tenotomy), or by passive stretch. In all these three conditions there is evidence that the hypertrophy is associated with altered distribution of either ribosomes or myosin mRNA.

Severance of the distal tendon of the gastrocnemius muscle causes a hypertrophy in the synergistic plantaris muscle and this is associated with a transient increase in the proportion of ribosomes in the myofibrillar cytoplasm as opposed to those found in the subsarcolemmal region (Horne & Hesketh, 1990*b*). Similarly, hypertrophy induced by the β-adrenoreceptor agonist clenbuterol was also accompanied by an increased ratio of myofibrillar: subsarcolemmal ribosomes (Horne & Hesketh, 1990*b*). In the latter instances the effect appeared to be rapid and occurred during the early stage of the hypertrophic response of the muscle when the increase in overall protein synthesis was due apparently to increases in both translation rates and total ribosomal content (Hesketh *et al.*, 1992). Under these conditions the amount of both actin or myosin heavy chain mRNA, expressed per unit or ribosomal RNA, was unchanged. Since total RNA (approximately 80% ribosomal RNA) was increased, the total amount of actin and myosin mRNA must have also been elevated. Thus, the greater proportion of ribosomes in the myofibrillar cytoplasm is probably due to increased targeting of polysomes containing the mRNAs for proteins such as actin and myosin in this compartment; since total RNA, and thus ribosomes, also increased it would appear that this involves targeting of newly synthesized mRNA/ribosomes rather than a redistribution of the existing protein synthetic apparatus. The altered ribosome distribution observed under conditions of increased synthesis of myofibrillar proteins is consistent with the hypothesis that the ribosomes in the intermyofibrillary cytoplasm are involved in the synthesis of the myofibrillar proteins.

Muscle fibres exhibit hypertrophy in response to stretch, and it has been known for some time that this is due to addition of sarcomeres at the ends of the fibres close to the myotendinous junction (Williams & Goldspink, 1973; Williams *et al.*, 1986). Rates of protein synthesis are greater towards the ends of normal fibres and are accelerated under stretch conditions (Williams *et al.*, 1986). This has recently been shown to be associated with an accumulation of myosin heavy chain mRNA close to the myotendinous junction, in a region where there is a large cytoplasmic space containing polysomes and developing myofibrils (Dix & Eisenberg, 1990). In addition, as shown in Fig. 3, the staining of individual fibres with anti-ribosomal antibodies reveals that, during stretch-induced hypertrophy, there is a particularly marked accumu-

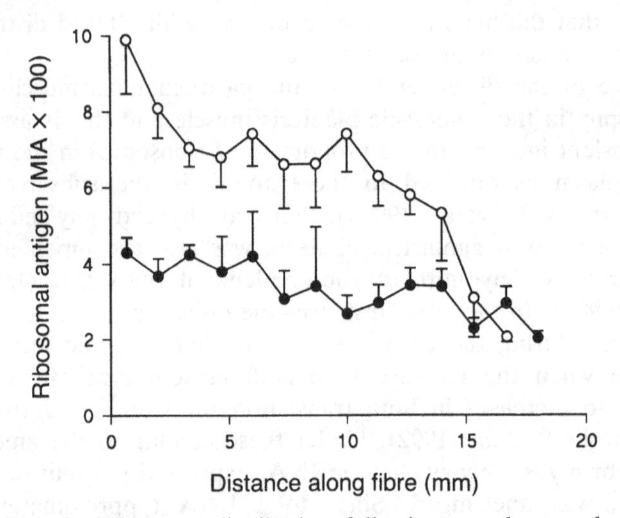

Fig. 3. Ribosome distribution following muscle stretch. Ribosome distribution was studied in individual fibres from rat anterior tibialis muscles which had undergone passive stretch (open circles) and from control fibres (closed circles). Distribution was assessed by immunohistochemistry using an anti-60S ribosomal subunit antibody (Horne & Hesketh, 1990*a*) and quantified by microdensitometry taking readings along the whole length of the fibres. Values are shown as means ± sem from three fibres from three animals. Distance was measured from the myotendinous junction. There was a statistically significant difference between the five readings at the proximal and distal ends of the fibres ($P < 0.05$) in the stretched muscle but not in the controls.

lation of ribosomal material close to the myotendinous junction. There is also increased ribosomal material in the mid-region of the fibre, compatible with the observed increase in myosin heavy chain mRNA in this area (Dix & Eisenberg, 1991; Russell *et al.*, 1992*a,b*). *In situ* hybridization studies have also shown that myosin heavy chain mRNA is redistributed in muscles that are undergoing repair (Russell *et al.*, 1992*a*), a situation in which there is also myofibril assembly in particular regions of the fibres and thus a requirement for newly synthesized proteins to be targeted to particular sites.

It is evident, therefore, that during growth, hypertrophy or repair of muscle, components of the protein synthetic apparatus are targeted to particular sub-cellular sites so that sarcometric proteins can be synthesized close to their site of function. Furthermore these processes are regulated according to the physiological need of the tissue.

mRNA localization and targeting: role of cytoskeleton and 3′ untranslated region

Subcellular mRNA localization was first observed in oocytes of *Xenopus* and *Drosophila* where certain specific mRNAs were found to be localized at one pole of the egg (Berleth *et al.*, 1988; Mowry & Melton, 1992). Hybridization studies *in situ* have since shown examples of mRNA localization in highly polarized cells such as intestinal epithelial cells (actin mRNA is localized in the apical region; Cheng & Bjerknes, 1989) and neurons (MAP2 mRNA in dendrites; Garner, Tucker & Matus, 1988). It is now also evident that mRNA localization occurs in cells which appear spatially less complicated and polarized; in spreading fibroblasts β-actin mRNA is located in the cell periphery close to the lamellipodia, a site of high actin protein concentration (Sundell & Singer, 1990). This localization is unaffected by puromycin and thus appears independent of nascent polypeptide chains. It is disrupted however by cytochalasin D, suggesting that microfilaments are required either for the transport or the anchoring of the mRNA (Sundell & Singer, 1991).

More recently it has been shown that, in myoblasts, the mRNAs coding for β-and γ-actin isoforms are localized in different cell domains with the β-form in the cell periphery and the γ-form in the perinuclear cytoplasm (Hill & Gunning, 1993). There is also evidence that vimentin and myosin heavy chain mRNAs (see above) and acetylcholine receptor subunit mRNA (Fontaine *et al.*, 1988) are localized to some extent in skeletal muscle. It would appear likely therefore that the redistribution of mRNAs and ribosomes during muscle growth and hypertrophy involves the directed transport of mRNAs.

mRNA localization in myofibres, and in cells in general, could theoretically arise by diffusion of the mRNA or the polyribosome complex, or by some specific transport mechanism, for example involving the cytoskeleton. Myosin heavy chain and other mRNAs have been observed to be concentrated around the nucleus (see Russell & Dix, 1992) and in the subsarcolemmal cytoplasm (see Fig. 2) as expected on a diffusion model. However, calculation of radial diffusion coefficients (Russell & Dix, 1992) gives values that are 2000-fold lower than expected for polysomes, suggesting that simple diffusion cannot account for the distribution of the mRNA–polysome complex.

In a variety of cells grown in culture, approximately 75% of mRNAs and polysomes are associated with the cell matrix (e.g. see Taneja *et al.*, 1992); however, this is due not only to an association of some polysomes with the cytoskeleton but also the presence of rough endoplasmic reticulum in the cell matrix (Hesketh & Pryme, 1991). Using salt treatment or cytochalasins it has proved possible to release the population of polysomes which were bound to microfilaments (Ramaekers *et al.*, 1983; Hesketh & Pryme, 1991; Vedeler, Pryme & Hesketh, 1991) and it has been estimated that some 25–40% of polysomes are associated with the cytoskeleton (cytoskeletal-bound polysomes, CBP) in cultured cells (Hesketh, 1994). Immunohistochemical data have indicated that initiation factors (Shestakova *et al.*, 1993), ribosomes (Toh *et al.*, 1980) and cap-binding protein (Zumbe, Staehli & Trachsel, 1982) co-distribute with cytoskeletal components. In summary, there is now a strong body of evidence that a proportion of polysomes are associated with the cytoskeleton (Hesketh, 1994).

It appears that specific mRNAs are associated with CBP (Bird & Sells, 1986; Hesketh *et al.*, 1991*b*; Hesketh, 1994), suggesting that CBP are involved in the synthesis of specific proteins. Studies of cells transfected with chimaeric gene constructs in which the 3′ untranslated regions (3′ UTR) of β-globin and *c-myc* mRNAs were exchanged (Hesketh *et al.*, 1994) have shown both that the association of the *c-myc* mRNA with CBP is dependent on the 3′UTR and that the *c-myc* 3′UTR can target the β-globin coding sequences from free to cytoskeletal-bound polysomes. Furthermore, exchanging the 3′UTRs altered the localization of the *c-myc* coding sequences from perinuclear to peripheral. These results, together with those from similar experiments with actin 3′UTR sequences (Kislauskis *et al.*, 1993), extend the data from oocytes which implicate 3′UTR sequences in mRNA localization. It appears that there is a universal mechanism of mRNA localization, in somatic cells as well as oocytes, in which the 3′UTRs of certain mRNAs and the cytoskeleton are involved in targeting of

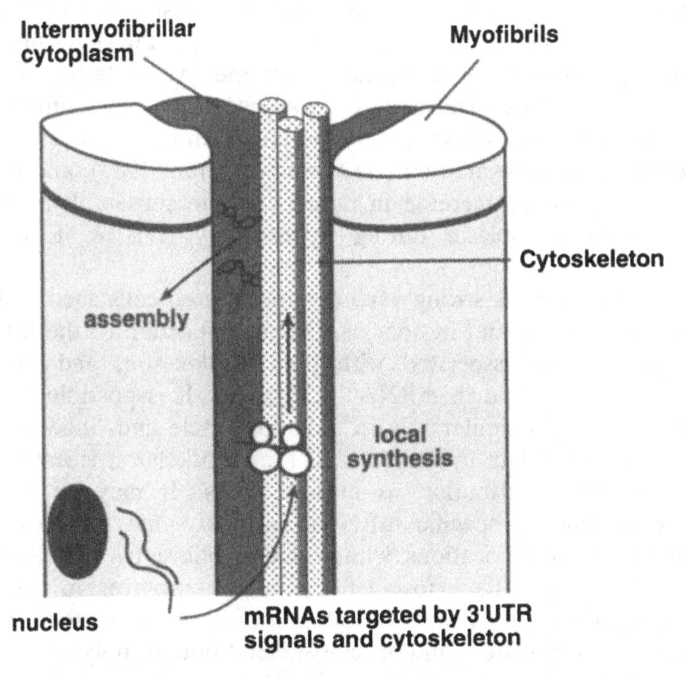

Fig. 4. mRNA targeting and local synthesis in muscle. A speculative scheme illustrating how the cytoskeleton and 3′untranslated regions of specific mRNAs may be involved in targeting of mRNAs and polysome complexes to the intermyofibrillary cytoplasm so as to achieve local synthesis of myofibrillar and costamere proteins.

those mRNAs for translation on cytoskeletal-bound polysomes in particular subcellular locations (Hesketh, 1994; Hesketh *et al.*, 1994).

The presence of cytoskeletal-bound polysomes in myoblasts (Bagchi, Larson & Sells, 1987; Bird & Sells, 1986) suggests that the association of mRNAs and polysomes with the cytoskeleton may be important in muscle fibres *in vivo*. As discussed above, in muscle fibres a proportion of polysomes are present within the intermyofribillary cytoplasm, and it has been suggested that such polysomes may be a particular case of 'cytoskeletal-bound polysomes' involved in the synthesis of myofibrillar proteins (Horne & Hesketh, 1990*a*). Furthermore, this hypothesis is supported by the observations that some mRNAs and polysomes are associated with the cytoskeleton in muscle fibres. Myosin heavy chain mRNA has been reported to be associated with cytoskeletal filaments in embryonic muscle (Pomeroy *et al.*, 1991) and in myocytes myosin heavy chains are synthesized on polysomes that are retained in the cell matrix and released following microfilament disruption induced by cytochalasin B (Isaacs & Fulton, 1987). In the Purkinje fibres of the heart, polysomes are associated with the cytoskeleton, possibly intermediate filaments (Thornell & Eriksson, 1981). In addition, the distribution of myosin heavy chain mRNA is similar to that of intermediate filaments in striated muscle (Craig & Pardo, 1983), and desmin intermediate filaments increase in number in the subsarcolemmal and intermyofibrillar cytoplasm during hypertrophy (Dix & Eisenberg, 1991).

In summary, there is strong evidence in cultured cells such as fibroblasts and myoblasts, and in oocytes, that a proportion of the mRNAs and polysomes are associated with the cytoskeleton, and that the cytoskeleton is involved in mRNA localization. It is possible that the cytoskeleton plays a similar role in striated muscle and, indeed, there is some correlative data to suggest that the cytoskeleton is involved in mRNA/polysome distribution in muscle fibres. It may be that, as illustrated in Fig. 4, specific mRNAs and polysome complexes are targeted to different locations within the myofibre by signals within the 3′UTR and via the cytoskeleton. Further progress in this area would be greatly aided by the development of methods for the isolation of myofibrillar-associated and/or cytoskeletal-bound polysomes from striated muscle. The identification of mRNA-binding proteins which also interact with the cytoskeleton or myofibrils will be an additional important step in characterization of the targeting mechanism. Crucially, such a targeting mechanism must be able to respond to the changing physiological needs of growth, repair and hypertrophy.

Differential nuclear activation

The muscle fibre is a long, multinucleated syncytium in which mRNA or protein targeting alone cannot account for changes in mRNA or ribosome distribution over large distances. For example, as shown in Fig. 3, in response to stretch there is an altered pattern of ribosome distribution along the length of the fibre and it would appear unlikely that this is due to retargeting of protein or mRNA from one end of the fibre to the other; alternatively, the data could be explained by a differential increase in ribosome synthesis in nuclear domains along the fibre.

Studies of cytosolic enzyme isoforms originally showed no evidence for nuclear domains in myofibres (see Hall & Ralston, 1989), and also suggested that the products of the different nuclei were mixed throughout the fibre. However, later data indicated that proteins destined for the Golgi apparatus or the myofibrils were localized close to their nucleus of origin (Pavlath *et al.*, 1989), indicating that targeting of these proteins led to restricted distribution. In contrast, it appears that cytosolic proteins are free to diffuse through the myoplasm and proteins on the cell surface free to diffuse through the membrane (Ralston & Hall, 1989), except in the case of N-CAM which is linked to the cytoskeleton. It is possible therefore that targeting of proteins themselves, or of their mRNAs (for example, to the cytoskeleton), restricts the protein distribution and leads to the formation of nuclear 'spheres of influence'. The existence of such nuclear domains is important because there is also evidence that myonuclei may differ in their gene expression as a result of extracellular factors. Thus, innervation results in local expression of a fast myosin isoform in a dually innervated fibre (Salviati, Biasia & Aloisi, 1986); furthermore, innervation also leads to a concentration of acetylcholine receptor mRNA in nuclei near the endplates (Fontaine *et al.*, 1988).

It would appear therefore that the altered profile of ribosome concentration (Fig. 3) and synthetic activity along the muscle fibre that one finds in stretch-induced hypertrophy is most likely due, at least in part, to a differential or selective increase in ribosome synthesis in the different nuclear domains, so giving rise to differences in the capacity for protein synthesis. Furthermore, such a differential activation of the nuclei along the fibre implies that there is an ability of the signalling mechanisms to elicit such differential responses. The unravelling of the control mechanisms involved awaits detailed knowledge of the signalling events, particularly the transcriptional events, involved.

c-myc expression in muscle hypertrophy

One regulatory transcriptional event which may be involved in hypertrophy is the expression of the cellular oncogene *c-myc*. Although *c-myc* is down-regulated during myogenesis, its mRNAs are present in low abundance in mature cardiac and skeletal muscle (Whitelaw & Hesketh, 1992). Furthermore, hypertrophy of both cardiac and skeletal muscle is associated with increased expression of *c-myc*: in the heart, a rapid increase in expression of *c-myc* mRNA (Mulvagh *et al.*, 1987; Izumo, Nadal-Ginard & Mahdavi, 1988) occurs 2 hours after imposition of a pressure overload; in skeletal muscle, hypertrophy induced either by the β-agonist clenbuterol or by the compensatory increase in workload following tenotomy also leads to a rapid, transient increase in expression of *c-myc* mRNA (Whitelaw & Hesketh, 1992).

A major question concerning these observations is whether the observed changes in *c-myc* occurred in myonuclei, and thus represent myonuclear activation and signalling events *within* the myofibre, in satellite cells, or in non-myogenic cells such as macrophages and fibroblasts. Although in cardiac hypertrophy there is evidence that the increased *c-myc* expression is largely outwith the myofibres (Snoeckx *et al.*, 1991; Hannan, Stennard & West, 1994), a number of arguments suggest that increased expression of *c-myc* can occur within the myonuclei. Firstly, agents which induce hypertrophy in cardiac myocytes in culture also induce *c-myc* expression (Starksen *et al.*, 1986). Secondly, the increase in *c-myc* expression observed after tenotomy or clenbuterol treatment (2–8 fold) is too great to be accounted for by increased expression only in satellite cells, which account for as little as 2–10% of the cell mass in mature muscle. Thirdly, *c-myc* expression can be induced in myotubes in culture (Endo & Nadal-Ginard, 1986). Lastly, preliminary immunocytochemistry data (Fig. 5) demonstrates that, after tenotomy, the frequency of nuclei which were both along the edges of the myofibres and stained by anti-*c-myc* antibody is such that it is improbable that the distribution can be wholly accounted for by satellite cells aligned outside the fibre. It is clearly vital to confirm these observations; however, at present, it appears that increased *c-myc* expression is part of the signalling mechanism by which the myofibre

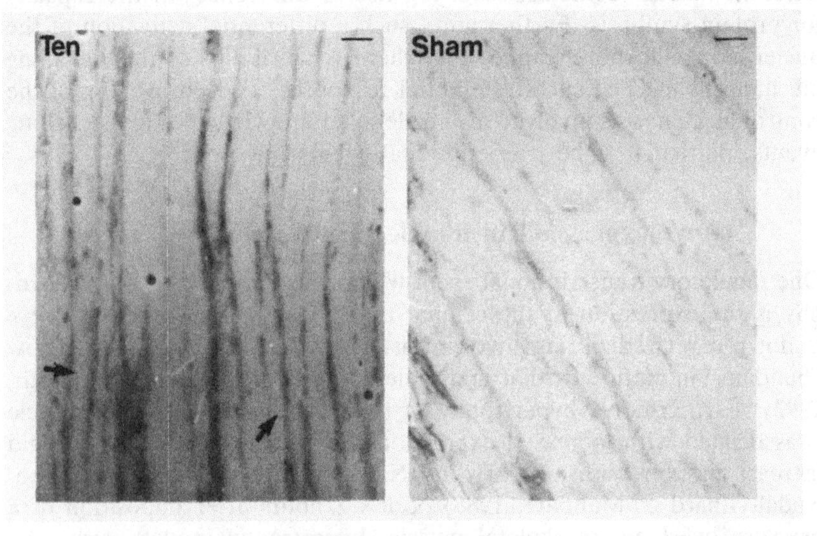

Fig. 5. C-myc distribution in hypertrophying muscle. C-myc protein distribution was studied by immunocytochemistry using a monoclonal anti-*c-myc* antibody and 10 μm sections from muscles undergoing hypertrophy after severance of the tendon to a synergistic muscle (ten) and from sham-operated controls (sham). Note the distribution of staining along the fibres (arrows). Bar = 25 μm.

responds to hypertrophic stimuli. It may be that *c-myc* will be a useful marker for activated myonuclei.

In *Xenopus* oocytes and in early embryonic development *c-myc* expression is not related to DNA synthesis (Godeau *et al.*, 1986) and it has been suggested that it may activate ribosome synthesis or regulate ribosomal RNA turnover (Gibson *et al.*, 1992). Muscle hypertrophy also represents a situation where there is ribosomal RNA accumulation in the absence of cell division, and the induction of *c-myc* may form part of the signal pathway which initiates the ribosome accumulation which is characteristic of muscle hypertrophy. It will be particularly interesting to investigate whether *c-myc* is induced in stretched muscle, and if so where the increased expression occurs – either along the fibre length or within specific domains, and whether the distribution corresponds to that of ribosome accumulation.

Summary

Since muscle fibres are structurally complex, highly organized syncytia containing many myonuclei, organization of the protein synthetic apparatus is crucial so that during hypertrophy or repair, the growth of fibres in length or in girth is controlled in an integrated manner. Hypertrophy involves not simply increased synthesis of mRNAs and ribosomes but accumulations in particular locations. There is increasing evidence that this targeting of newly synthesized proteins to the appropriate sites in the fibre involves both selective activation of different nuclei and local synthesis of myofibrillar proteins close to their site of function. Such spatial organization of the protein synthetic apparatus, which can respond to physiological needs, may involve mRNA targeting via the cytoskeleton. There is, in addition, evidence that there are changes in *c-myc* expression during muscle hypertrophy. *C-myc* appears to be induced as an early event in muscle hypertrophy, possibly as part of a signal pathway leading to activation of ribosome accumulation.

Acknowledgements

The author's work is supported by the Scottish Office Agriculture, Environment and Fisheries Department (SOAEFD).

References

Aigner, S. & Pette, D. (1990). *In situ* hybridization of slow myosin heavy chain mRNA in normal and transforming rabbit muscles with the use of a nonradioactively labeled cRNA. *Histochemistry*, **95**, 11–18.

Bagchi, T., Larson, D.E. & Sells, B.H. (1987). Cytoskeletal association of muscle-specific mRNAs in differentiating L6 rat myoblasts. *Experimental Cell Research*, **168**, 160–72.

Berleth, T., Burri, M., Thoma, G., Bopp, D., Richstein, S., Frigerio, G., Noll, M. & Nusslein-Volhard, C. (1988). The role of localization of bicoid RNA in organizing the anterior pattern of the Drosophila embryo. *EMBO Journal*, **7**, 1749–56.

Bird, R.C. & Sells, B.H. (1986). Cytoskeleton involvement in the distribution of mRNP complexes and small cytoplasmic RNAs. *Biochemica et Biophysica Acta*, **868**, 215–25.

Cheng, H. & Bjerknes, M. (1989). Asymmetric distribution of actin mRNA and cytoskeletal pattern generation in polarized epithelial cells. *Journal of Molecular Biology*, **210**, 541–9.

Craig, S.W. & Pardo, J.V. (1983). Gamma actin, spectrin and intermediate filament proteins colocalize with vinculin at costameres, myofibril-t-sarcolemma attachment sites. *Cell Molility*, **3**, 449–62.

Cripe, L., Morris, E. & Fulton, A.B. (1993). Vimentin mRNA location changes during muscle development. *Proceedings of the National Academy of Sciences, USA*, **90**, 2724–8.

Dix, D.J. & Eisenberg, B.R. (1988). *In situ* hybridization and immunocytochemistry in serial sections of rabbit skeletal muscle to detect myosin expression. *Journal of Histochemistry and Cytochemistry*, **6**, 1519–26.

Dix, D.J. & Eisenberg, B.R. (1990). Myosin mRNA accumulation and myofibrillogenesis at the myotendinous junction of stretched muscle fibres. *Journal of Cell Biology*, **111**, 1885–94.

Dix, D.J. & Eisenberg, B.R. (1991). Redistribution of myosin heavy chain mRNA in the midregion of stretched muscle fibres. *Cell Tissue Research*, **263**, 61–9.

Endo, T. & Nadal-Ginard, B. (1986). Transcriptional and posttranscriptional control of c-myc during myogenesis: its mRNA remains inducible in differentiated cells and does not suppress the differentiated phenotype. *Molecular and Cellular Biology*, **6**, 1412–21.

Fontaine, B., Sassoon, D., Buckingham, M. & Changeux, J.P. (1988). Detection of the nicotinic acetylcholine receptor alpha-subunit mRNA by *in situ* hybridization at neuromuscular junctions of 15-day-old chick striated muscles. *EMBO Journal*, **7**, 603–9.

Galavazi, G. (1971). Identification of helical polyribosomes in sections of mature skeletal muscle fibres. *Zeitschrift für Zellforschung und Microskopische Anatomie*, **121**, 531–47.

Garner, C.C., Tucker, R.P. & Matus, A. (1988). Selective localization of messenger RNA for cytoskeletal protein MAP2 in dendrites. *Nature*, **336**, 674–7.

Gauthier, G.F. & Mason-Savas, A. (1993). Ribosomes in the skeletal muscle filament lattice. *Anatomical Records*, **237**, 149–56.

Gauthier, G.F. & Schaeffer, S.F. (1974). Ultrastructural and cytochemical manifestations of protein synthesis in the peripheral sarcoplasm of denervated and newborn skeletal muscle fibres. *Journal of Cell Science*, **14**, 113–37.

Gibson, A.W., Ye, R., Johnston, R.N. & Broxder, L.W. (1992). A possible role for *c-myc* oncoproteins in post-transcriptional regulation of ribosomal RNA. *Oncogene*, **7**, 2363–7.

Godeau, F., Persson, H., Gray, H.E. & Pardee, A.B. (1986). *c-myc* expression is dissociated from DNA synthesis and cell division in *Xenopus* oocytes and early development. *EMBO Journal*, **5**, 3571–7.

Hall, Z.W. & Ralston, E. (1989). Nuclear domains in muscle cells. *Cell*, **59**, 771–2.

Hannan, R.D., Stennard, F.A. & West, A.K. (1994). Localization of *c-myc* protooncogene expression in rat heart *in vivo* and in the isolate, perfused heart following treatment with norepinephrine. *Biochemica et Biophysica Acta*, **1217**, 281–90.

Hesketh, J.E. (1994). Translation on the cytoskeleton–a mechanism for targeted protein synthesis. *Molecular Biology Reports*, **19**, 233–43.

Hesketh, J.E. & Pryme, I.F. (1991). Interaction between mRNA, ribosomes and the cytoskeleton. *Biochemical Journal*, **277**, 1–10.

Hesketh, J.E., Campbell, G.P., Lobley, G.E., Maltin, C.A., Acamovic, F. & Palmer, R.M. (1992). Stimulation of actin and myosin synthesis in rat gastrocnemius muscle by clenbuterol; evidence for translational control. *Comparative Biochemistry and Physiology*, **102C**, 23–7.

Hesketh, J.E., Campbell, G.P. & Loveridge, N. (1991*a*). Myosin heavy chain mRNA is present in both myofibrillar and subsarcolemmal regions of muscle fibres. *Biochemistry Journal*, **279**, 309–10.

Hesketh, J.E., Campbell, G.P. & Whitelaw, P.F. (1991*b*). *C-myc* messenger RNA in cytoskeletal-bound polysomes in fibroblasts. *Biochemical Journal*, **274**, 607–9.

Hesketh, J.E., Campbell, G.P., Piechaczyk, M. & Blanchard, J.-M. (1994). Targeting of *c-myc* and β-globin coding sequences to cytoskeletal-bound polysomes by *c-myc* 3′ untranslated region. *Biochemical Journal*, **298**, 143–8.

Hill, M.A. & Gunning, P. (1993). Beta and gamma actin mRNAs are differentially located in myoblasts. *Journal of Cell Biology*, **122**, 825–32.

Horne, Z. & Hesketh, J. (1990*a*). Immunological localization of ribosomes in striated rat muscle. Evidence for myofibrillar association and ontological changes in the subsarcolemmal:myofibrillar distribution. *Biochemical Journal*, **268**, 231–6.

Horne, Z. & Hesketh, J. (1990*b*). Increased association of ribosomes with myofibrils during the skeletal-muscle hypertrophy induced either by the beta-adrenoiceptor agonist clenbuterol or by tenotomy. *Biochemical Journal*, **272**, 831–3.

Isaacs, W. & Fulton, A.B. (1987). Cotranslational assembly of myosin heavy chain in developing cultured skeletal muscle. *Proceedings of the National Academy of Sciences, USA*, **84**, 6174–8.

Izumo, S., Nadal-Ginard, B. & Mahdavi, V. (1988). Proto-oncogene induction and reprogramming of cardiac gene expression produced by pressure overload. *Proceedings of the National Academy of Sciences, USA*, **85**, 339–43.

Kislauskis, E.H., Li, Z., Taneja, K.L. & Singer, R.H. (1993). Isoform-specific 3′ untranslated sequences sort α-cardiac and β-cytoplasmic actin messenger RNAs to different cytoplasmic compartments. *Journal of Cell Biology*, **123**, 165–72.

Larsen, P.F., Hudgson, P. & Walton, J.N. (1969). Morphological relationship of polyribosomes and myosin filaments in developing and regenerating skeletal muscle. *Nature*, **222**, 1168–9.

Larsen, T.H., Hesketh, J.E., Rotevatn, S., Greve, G. & Saetersdal, T. (1994). Ribosome distribution in normal and infarcted rat hearts. *Histochemistry Journal*, **26**, 79–89.

Lawrence, J.B. & Singer, R.H. (1986). Intracellular localization of mRNAs for cytoskeletal proteins. *Cell*, **45**, 407–15.

Lewis, S.E.M., Kelly, F.J. & Goldspink, D.F. (1984). Pre- and post-natal growth and protein turnover in smooth muscle, heart and slow- and fast-twitch skeletal muscles of the rat. *Biochemical Journal*, **217**, 517–26.

Morris, E.J. & Fulton, A.B. (1994). Rearrangement of mRNAs for costamere proteins during costamere development in cultured skeletal muscle from chicken. *Journal of Cell Science*, **107**, 377–86.

Mowry, K.L. & Melton, D.A. (1992). Vegetal messenger RNA localization directed by a 340 nt RNA sequence element in *Xenopus* oocytes. *Science*, (Wash. DC) **255**, 991–4.

Mulvagh, S.L., Michael, L.H., Perryman, M.B., Roberts, R. & Schneider, M.D. (1987). A hemodynamic load *in vivo* induces cardiac expression of the cellular oncogene, *c-myc*. *Biochemical and Biophysical Research Communications*, **147**, 627–36.

Pavlath, G.K., Rich, K., Webster, S.G. & Blau, H.M. (1989). Localization of muscle gene products in nuclear domains. *Nature*, **337**, 570–3.

Pomeroy, M.E., Lawrence, J.B., Singer, R.H. & Billings-Gagliardi, S. (1991). Distribution of myosin heavy chain mRNA in embryonic muscle tissue visualized by ultrastructural *in situ* hybridization. *Developmental Biology*, **143**, 58–67.

Ralston, E. & Hall, Z.W. (1989). Intracellular and surface distribution of a membrane protein (CD8) derived from a single nucleus in multinucleated myofibres. *Journal of Cell Biology*, **109**, 2345–52.

Ramaekers, F., Benedetti, E., Dunia, I., Vorstenbosch, P. & Bloemendal, H. (1983). Polyribosomes associated with microfilaments in cultured lens cells. *Biochemica et Biophysica Acta*, **740**, 441–8.

Russell, B. & Dix, D.J. (1992). Mechanisms for intracellular distribution of mRNA: *in situ* hybridization studies in muscle. *American Journal of Physiology*, **262**, C1–C8.

Russell, B., Dix, D.J., Haller, D.L. & Jacobs-El, J. (1992*a*). Repair of injured skeletal muscle: a molecular approach. *Medical Science Sports Exercise*, **24**, 189–96.

Russell, B., Wenderoth, M.P. & Goldspink, P.H. (1992*b*). Remodelling of myofibrils: subcellular distribution of myosin heavy chain mRNA and protein. *American Journal of Physiology*, **262**, R339–45.

Salviati, G., Biasia, E. & Aloisi, M. (1986). Synthesis of fast myosin induced by fast ectopic innervation of rat soleus muscle is restricted to the ectopic endplate region. *Nature*, **322**, 637–9.

Shestakova, E.A., Motuz, L.P., Minin, A.A. & Gavrilova, L.P. (1993). Study of localization of the protein-synthesizing machinery along actin filament bundles. *Cell Biology International*, **17**, 409–16.

Snoeckx, L.H.E.H., Contard, F., Samuel, J.L., Marotte, F. & Rappaport, L. (1991). Expression and cellular distribution of heat-shock and nuclear oncogene proteins in rat hearts. *American Journal of Physiology*, **261**, H1443–51.

Starksen, N.F., Simpson, P.C., Bishopric, N., Coughlin, S.R., Lee, W.M.F., Excobedo, J.A. & Williams, L.T. (1986). Cardiac myocyte hypertrophy is associated with c-myc proto-oncogene expression. *Proceedings of National Academy of Sciences. USA*, **83**, 8348–50.

Sundell, C.L. & Singer, R.H. (1990). Actin mRNA localizes in the absence of protein synthesis. *Journal of Cell Biology*, **111**, 2397–403.

Sundell, C.L. & Singer, R.H. (1991). Requirement of microfilaments in sorting of actin messenger RNA. *Science*, **253**, 1275–7.

Taneja, K.L., Lifshitz, L.M., Fay, F.S. & Singer, R.H. (1992). Poly(A) RNA codistribution with microfilaments: evaluation by *in situ* hybridization and quantitative digital imaging microscopy. *Journal of Cell Biology*, **119**, 11245–60.

Thornell, L.-E, & Eriksson, A. (1981). Filament systems of the Purkinje fibres of the heart. *American Journal of Physiology*, **241**, H291–305.

Toh, B., Lolait, S., Mathy, J. & Baum, R. (1980). Association of mitochondria with intermediate filaments and of polyribosomes with cytoplasmic actin. *Cell Tissue Research*, **211**, 163–9.

Vedeler, A., Pryme, I.F. & Hesketh, J.E. (1991). The characterization of free, cytoskeletal and membrane-bound polysomes in Krebs II ascites and 3T3 cells. *Molecular and Cellular Biochemistry*, **100**, 183–93.

Waterlow, J.C., Garlick, P.J. & Millward, D.J. (1978). *Protein Turnover in Mammalian Tissues and in the Whole Body*, pp. 625–695. Oxford: Elsevier/North-Holland Publishing Co.

Whitelaw, P.F. & Hesketh, J.E. (1992). Expression of *c-myc* and *c-fos* is rat skeletal muscle: evidence for increased levels of *c-myc* mRNA during hypertrophy. *Biochemical Journal*, **281**, 143–7.

Williams, P.E. & Goldspink, G. (1973). The effect of immobilization on the longitudinal growth of striated muscle fibres. *Journal of Anatomy*, **116**, 45–55.

Williams, P., Watt, P., Bicik, V. & Goldspink, G. (1986). Effect of stretch combined with electrical stimulation on the type of sarcomeres produced at the ends of muscle fibres. *Experimental Neurology*, **93**, 500–9.

Zumbe, A., Staehli, C. & Trachsel, H. (1982). Association of a 50 K molecular weight cap binding protein with the cytoskeleton in baby hamster cells. *Proceedings of the National Academy of Sciences, USA*, **79**, 2927–31.

P.T. LOUGHNA and C. BROWNSON

The role of mechanical tension in regulating muscle growth and phenotype

Introduction

Cells in the body are subjected to both static and dynamic physical forces. In the musculoskeletal system, cells are constantly under the stress of gravity, as well as to tension resulting from muscular contractions of widely varying intensities. Hypertrophy of cardiac, skeletal and arterial smooth muscle and increased bone density are commonly described adaptation to increased mechanical stress. Interest in the mechanical load is transduced not only into such elevated cellular growth but also into modulation of phenotype in a variety of cell types has increased greatly over the last few years as indicated by increasing numbers of articles and reviews (Komuro & Yazaki, 1993; Morgan & Baker, 1991; Watson, 1991; Davies, 1995). The role of mechanical strain in the determination of cardiac muscle phenotype has received considerable attention, but the effect on skeletal muscle is less well documented. This may be due, in part, to a traditional emphasis on neural regulation of skeletal muscle phenotype. It is clear that both active and passive mechanical forces can induce skeletal muscle enlargement and this increase in the adult animal is due solely to fibre hypertrophy. This is a product of elevated protein synthesis which outweighs parallel increases in protein degradation (Loughna, Goldspink & Goldspink, 1986).

There are a number of *in vivo* and *in vitro* models that have been employed to investigate the actions of active and passive tension upon skeletal muscle including tenotomy, pinning of joints, centrifugation and weights attached to avian wings (Alway *et al.*, 1989; Morgan & Loughna, 1989; Sola, Christensen & Martin, 1973). The model employed in the studies described here is simple and non-invasive and involves immobilizing the limb, such that the muscles to be examined are held at greater than resting length. Conversely, muscles have also been immobilized at less than resting length, such that they experience reduced levels of tension and undergo atrophy (i.e. disuse atrophy). We have compared the effects of immobilization in these two positions upon the phenotype of fast phasic and slow postural muscles.

Modulation of phenotype

The contractile and metabolic properties of a muscle fibre are direct products of the contractile protein isoform composition of its sarcomeres, the relative levels of various enzymes in the cytoplasm and the functioning of the excitation–contraction coupling system. Each of the contractile protein components of the sarcomeres has a number of isoforms which are products of different genes and/or differential post-transcriptional processing of the primary transcript (for review see Swyngehdauw, 1986). We have examined the effects of immobilization in the lengthened position (passive stretch) and shortened position (disuse) upon the transcript levels of a number of contractile protein genes.

Skeletal muscle myosin heavy chain (MyHC) is the major component of the thick filament of the sarcomere and in the mammal exists as a number of isoforms, each of which is coded for by an individual member of a multigene family. These isoforms are expressed in a developmentally regulated manner and differ markedly between adult muscles of varying fibre type composition contributing to their differing speeds of contraction (Mahadavi *et al.*, 1986). The level of expression of MyHC isoforms is not static in different fibre types, but can be modulated by a variety of stimuli such as thyroid hormone, growth hormone and altered patterns of innervation, (Izumo, Nadal-Ginard & Mahadavi, 1986; Pette & Dusterhoft, 1992; Brownson *et al.*, 1992; Loughna & Bates, 1994). We have found that passive stretch also modulates the levels of MyHC transcripts in skeletal muscles (Loughna *et al.*, 1990). In the soleus, a slow postural muscle which normally contains very low levels of the IIB MyHC mRNA, there is a massive increase in levels of this transcript in response to disuse alone which is negated by passive stretch (Fig. 1). In contrast, in the fast phasic plantaris and mixed gastrocnemius muscles, where this transcript is present at high levels, disuse had no significant effect but passive stretch produced a reduction in these levels (Fig. 1). However, in these muscles disuse reduces the level of both fast type IIA and slow type I MyHC transcripts and passive stretch either prevents this reduction or in some cases increases it above control levels (Fig. 1). In the soleus the converse is true: disuse has little effect, but passive stretch leads to a reduction of these mRNAs. Thus it appears that immobilization in the

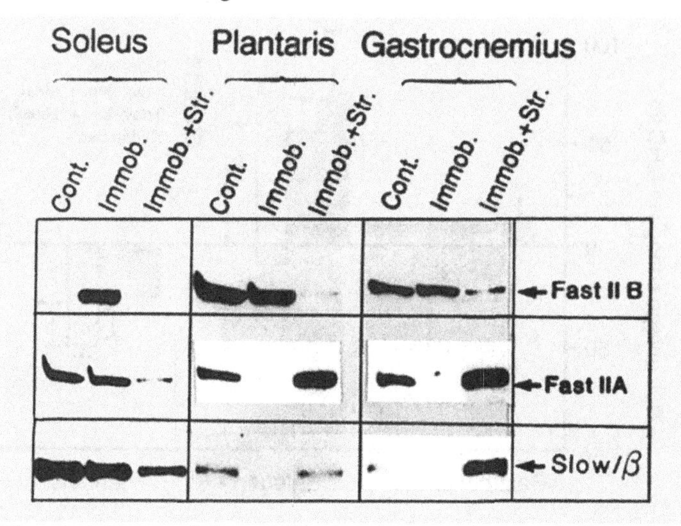

Fig. 1. Effects of immobilization in the shortened (Immob.) and lengthened position (Immob + str) upon mRNA levels of three adult myosin heavy chain isoforms in three different rat hindlimb muscles. The cDNA probes used were isoform specific and S1 nuclease digestion to total muscle RNA was carried out, following hybridization to total muscle (data from Loughna *et al.*, 1990).

shortened position has markedly different and often opposite effects upon the levels of an adult sarcomeric MyHC transcript to those produced by immobilization in the lengthened position. Each isoform responds differently to the same stimulus in different muscles and type IIB MyHC was modulated differentially from the other two adult isoforms.

In these *in vivo* models of passive stretch and disuse, it is not possible to disassociate changes in muscle tension induced by immobilization from possible effects to these procedures on activity patterns of the motor nerve supply. In order to distinguish between these two possibilities, the effects of immobilization in the disuse and stretch positions of muscles that had been previously denervated were also investigated. Denervation alone produced a dramatic elevation in the type IIB transcript in the soleus muscle similar to that observed in response to disuse in the innervated muscle (Fig. 2). Passive stretch prevented the elevation of this transcript in the denervated soleus demonstrating that tension plays a crucial role in the regulation of MyHC gene expression and this is independent of innervation.

Myosin heavy chain is the major component of the thick filament

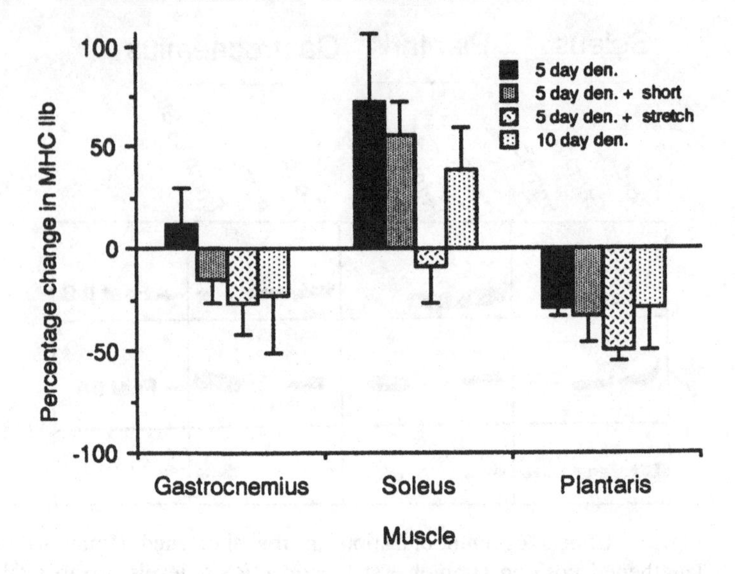

Fig. 2. Changes in fast glycolytic (type IIb) myosin heavy chain mRNA levels in three rat hindlimb muscles following denervation alone (den) and denervation with immobilization in either the shortened (den + short) or lengthened position (den + stretch). Denervation was produced by resectioning of the right sciatic nerve in the mid-thigh region under anaesthesia. The animals were divided into four groups in two of which the denervated limb was immobilized as described above and the animals were killed after 5 days. In the other two groups, the limbs were not immobilized and the animals killed after 5 and 10 days. Total RNA was extracted from the muscles, and the relative levels of type IIb myosin heavy chain were assessed by a standard slot blot procedure using a radioactively end-labelled isoform specific oligonucleotide probe. The levels of this transcript were elevated in the slow postural muscle, the soleus, but not in the other two muscles. This elevation in the soleus was prevented by passive stretch.

and of undoubted physiological significance. However, other sarcomeric protein components contribute to contraction and its regulation and all have a variety of isoforms. We have examined the effects of disuse and passive stretch upon the mRNA levels of a number of contractile proteins and have found that some are less responsive to these treatments than the MyHC gene products. For example, fast troponin T (TNT_f) mRNA levels are not modified dramatically following immobilization in either position, although there is a small reduction in the stretched plantaris (Fig. 3). The fast MLC transcript is similarly only

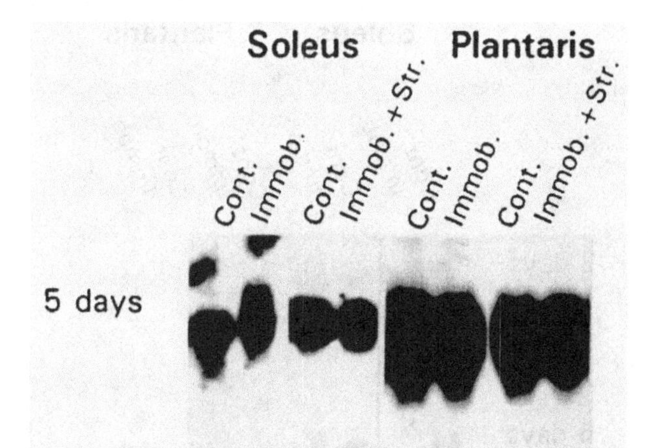

Fig. 3. Northern blot analysis of fast troponin T mRNA levels in the slow postural soleus and fast phasic plantaris muscles after 5 days of immobilization in the shortened (Immob) or lengthened position (Immob + str).

affected in the plantaris where it is unaffected by disuse but reduced by stretch (data not shown).

Proteins, particularly metabolic enzymes, which are not structural components of the sarcomere also contribute to muscle fibre phenotype. Such enzymes often do not exhibit the same extensive isoform diversity but are differentially expressed in different fibre types. It is unclear, however, whether such enzymes demonstrate similar differential and muscle-specific responses to disuse and passive stretch as those shown by members of the MyHC gene family. We have examined two such metabolic enzymes, carbonic anhydrase III (CAIII) which is normally predominant in type I fibres although it is dramatically induced in fast muscle subjected to chronic electrical stimulation (Brownson *et al.*, 1988) and phosphoglucoisomerase (PGI), a glycolytic enzyme more prevalent in type II fibres. CAIII mRNA levels, normally high in soleus, fell dramatically in this muscle in response to disuse (Fig. 4). The rapid disappearance of this transcript in the slow soleus muscle in response to disuse suggests that tension may play a role in its expression in skeletal muscle. The effects of stretch, upon soleus CAIII mRNA levels, were slightly more complex (Fig. 4). Whereas at 5 days, levels of this transcript were below control levels, the converse was true at an earlier 2-day timepoint. This suggests that there was an initial rise in response to the stretch stimulus which soon disappeared, perhaps due to a decline in the efficacy of the stretch response and/

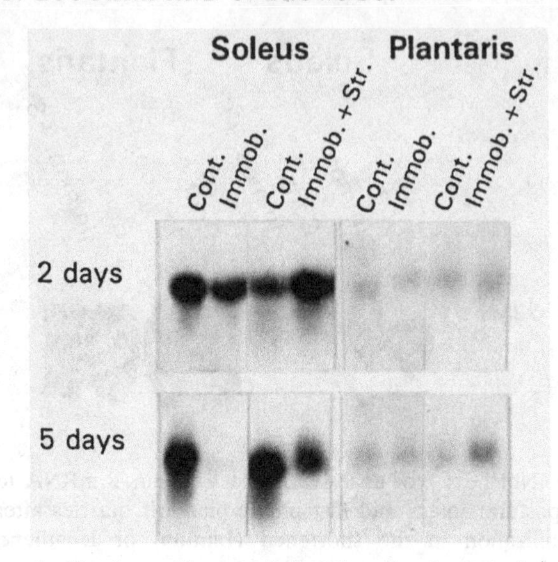

Fig. 4. Northern blot analysis of carbonic anhydrase III (CAIII) mRNA levels in the soleus and plantaris muscles after 2 and 5 days of immobilization in the shortened (Immob) and lengthened (Immob + str) position.

or the rapid turnover of this transcript. In the fast plantaris muscle CAIII mRNA levels were very low and were little affected by immobilization in either position. Disuse and passive stretch produced similar effects on PGI mRNA levels in both the soleus and the plantaris muscles. After 5 days of immobilization, the levels of the PGI transcript, in both muscles, were unaffected by disuse but were reduced in response to passive stretch (Fig. 5), suggesting a stretch-induced move to a more oxidative phenotype. These data demonstrate that the expression of soluble metabolic enzyme transcripts in muscle can be affected differently by disuse and passive stretch and that these effects occur in a muscle-specific manner.

Induction of fetal genes

Cardiac muscle responds to passive stretch in a manner similar to that observed in skeletal muscle in that it undergoes hypertrophy and modulation in phenotype associated with specific alterations in gene expression. One phenomenon of stretch induced adult cardiac muscle hypertrophy that has been noted in a number of animal models is the reinduction of what has been termed a 'fetal programme' (for review

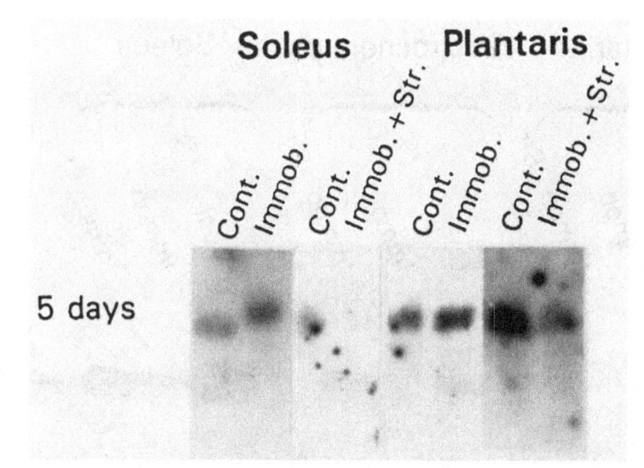

Fig. 5. Northern blot analysis of phosphoglucoisomerase (PGI) mRNA levels in the soleus and plantaris muscles following 5 days of immobilization in the shortened (Immob) and lengthened position (Immob+str).

see Parker & Schneider, 1991). This involves the rapid induction of genes, in adult ventricular muscle, that are normally only expressed at high levels in this tissue in the fetus, and which are down-regulated post-natally. In the rat, these include β-MyHC, skeletal α-actin and the atrial naturietic factor (Schwartz *et al.*, 1986; Izumo *et al.*, 1987; Izumo, Nadal-Ginard & Mahadavi, 1988). We examined levels of embryonic MyHC and cardiac α-actin transcripts, which in skeletal muscle are only expressed at high levels prenatally, in adult muscles which had been subjected to passive stretch by immobilization in the lengthened position. Passive stretch caused an up-regulation of the embryonic MyHC gene in all adult muscles examined, whereas disuse had no effect (Fig. 6). A similar up-regulation of embryonic MyHC has been observed in the wing-weighted model of avian skeletal muscle hypertrophy in the anterior latissimus dorsi (ALD).

On the other hand, cardiac α-actin mRNA, present in all adult muscles examined, was reduced in each case in response to disuse, and passive stretch partially or wholly prevented this reduction (Fig. 7). Fast phasic plantaris muscle has much lower levels of this transcript than the slow postural soleus, suggesting that expression of this transcript may be particularly sensitive to tension as the latter muscle is exposed continually to elevated tension levels. The tension placed on the muscles in this *in vivo* model is not great and is transient, and it is probable that a greater tension would produce elevated levels of this transcript. Thus, in contrast to adult contractile protein isoforms,

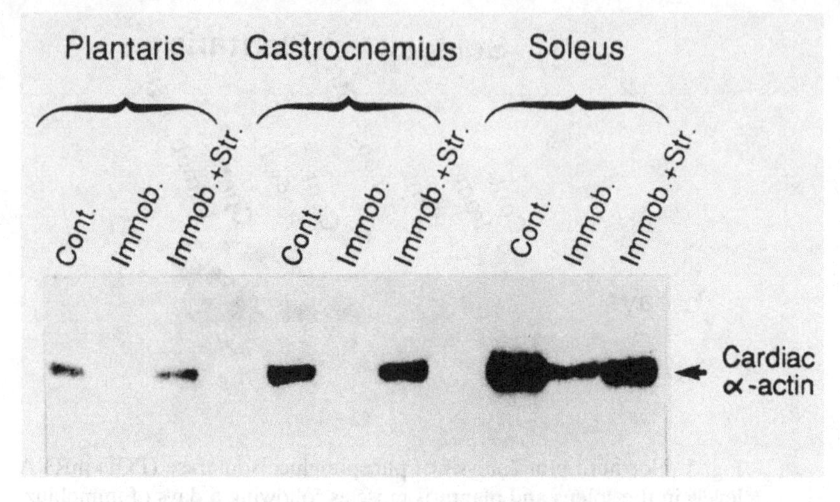

Fig. 6. Changes in cardiac α-actin mRNA levels, measured by S1 nucleus analysis, in three skeletal muscles following 5 days of immobilization in the shortened (Immob) and lengthened (Immob + str) position. The cDNA probe used was a single standard 75 nt fragment containing 20 nt of carboxy terminal coding sequence and 55 nt of the 3′ untranslated region of the rat cardiac α-actin transcript (from Loughna *et al.*, 1990).

which generally respond to stretch in a highly muscle-specific muscle manner, developmental isoforms appear to be actively up-regulated by mechanical tension in all muscle types. Skeletal muscle may therefore be subject to induction of a 'fetal programme', in response to stretch, similar to that which has been extensively characterized in cardiac muscle.

Two other possible explanations, however, could account for the up-regulation of developmental isoforms in adult muscle. First, the passive stretch employed in this model may cause damage to muscle fibres and the expression of developmental isoforms of contractile proteins may be associated with the regeneration of these fibres. Histological studies that we have carried out have shown no muscle damage and, in longer-term studies, we have not observed the centrally located nuclei associated with muscle fibre regeneration. Secondly, it has been suggested that the satellite cells that supply a hypertrophying muscle fibre with additional nuclei might be responsible solely for the expression of developmental isoforms. Recently, however, a study by McCormick and Schultz (1994) examined the relationship between satellite cells and embryonic myosin expression in the wing-weighted

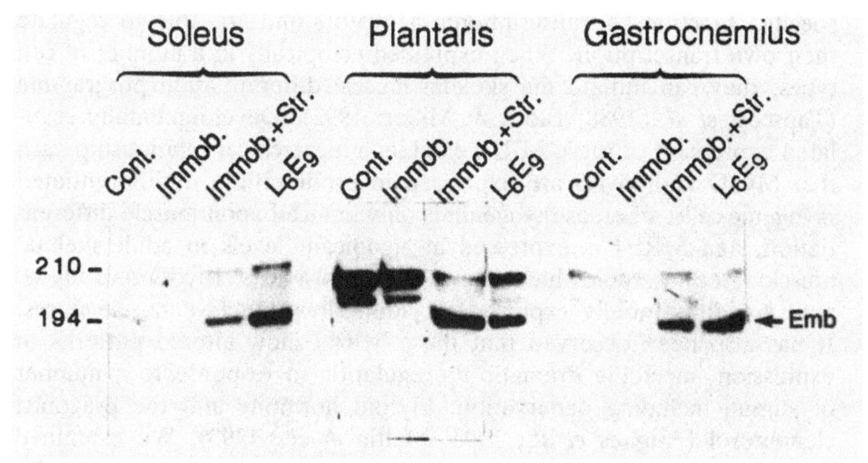

Fig. 7. Re-expression of the embryonic skeletal myosin heavy chain in adult muscle. For detection of this transcript, a 210 nt cDNA fragment was used for S1 nuclease mapping. This probe generates a fully protected fragment of 194 nt. The levels of this transcript were measured after 5 days of immobilization in the shortened (Immob) or lengthened (Immob+str) position. L6E9 differentiated myotube mRNA was used as a positive control for this transcript (from Loughna *et al.*, 1990).

model of hypertrophy in the avian ALD muscle. They observed that both endemic nuclei and newly incorporated satellite cell nuclei contributed to embryonic MyHC production in adult muscle fibres. At present, the available data suggests that the up-regulation of developmental isoforms, in response to passive stretch, is a general phenomenon occurring throughout the muscle fibre syncitium.

Possible role of myogenic regulatory factors

A number of studies suggest that adaptive changes in adult skeletal muscle are a result of modulation at the transcriptional level but the mediators involved in the transduction pathways between external stimuli and altered muscle gene expression have yet to be identified. Over the past few years there has been intensive investigation of a family of transcription factors known as the myogenic regulatory factors (MRFs). The gene products of this family have common structural motifs, the basic and helix–loop–helix domains, required for DNA binding and the consequent activation of a number of muscle-specific genes (Olsen, 1990; Davis *et al.*, 1990). They are skeletal muscle specific, function as transcriptional activators and are able to regulate their own transcription. When expressed ectopically in a number of cell types, they can initiate the skeletal muscle differentiation programme (Tapscott *et al.*, 1988; Block & Miller, 1992). Developmentally regulated expression of these MRFs exhibits a hierarchical relationship such that MyoD and Myf-5 are expressed in proliferating, undifferentiated myogenic cells, whereas myogenin is only induced upon muscle differentiation, and MRF4 is expressed at significant levels in adult skeletal muscle. Recent studies have shown that two MRFs, MyoD and myogenin are differentially expressed in adult slow and fast muscle fibres. It has also been observed that these MRFs show altered patterns of expression, including dramatic up-regulation in response to a number of stimuli including denervation, thyroid hormone and the β-agonist clenbuterol (Hughes *et al.*, 1993; Maltin *et al.*, 1993). We examined the effects of passive stretch upon the level of MRF transcripts to determine if any exhibited a general pattern of up-regulation as part of a stretch induced 'embryonic programme' as described above. Myogenin mRNA levels were not altered significantly by muscle stretch in the soleus but in the plantaris this transcript was elevated dramatically (Fig. 8). Disuse had little effect in either muscle. In contrast, MRF4 was up-regulated by stretch in the soleus but decreased in the plantaris (Fig. 9). Disuse had little effect on the levels of this transcript in the plantaris, but induced a marked reduction in the soleus.

The most dramatic response of a MRF transcript to passive stretch was the large elevation of myogenin mRNA levels in plantaris (Fig. 8). This transcript was unaffected by passive stretch in soleus muscle

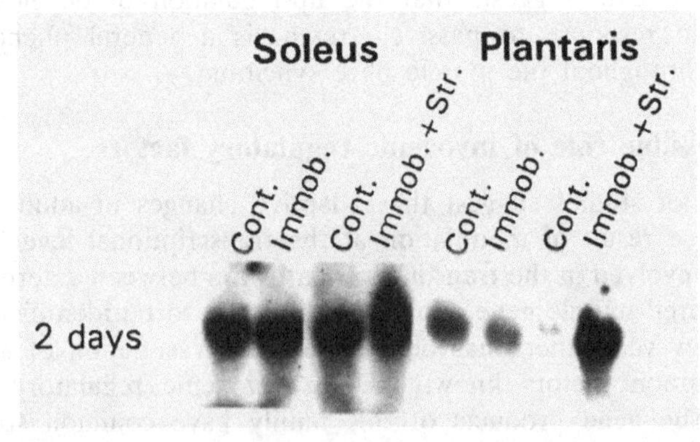

Fig. 8. Northern analysis of levels of myogenin mRNA in the soleus and plantaris following 2 days of immobilization in the shortened (Immob) and lengthened position (Immob + str).

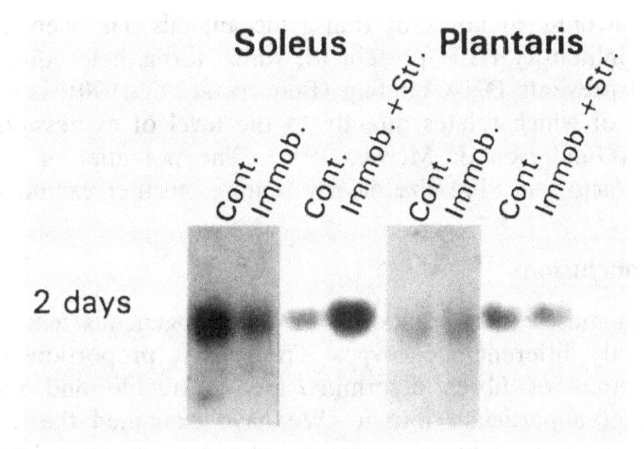

Fig. 9. Northern analysis of MRF4 mRNA levels in the soleus and plantaris following 2 days of immobilization in the shortened (Immob) and lengthened position (Immob + str).

where it is normally present at a higher level than in plantaris muscle. In contrast, MRF4 mRNA levels, which were down-regulated during disuse atrophy in soleus muscle, were up-regulated by the anabolic stretch stimulus in soleus muscle (Fig. 9). Passive stretch produced the opposite effect in plantaris muscle, clearly demonstrating a muscle-specific response in both myogenin and MRF4 mRNA levels to this stimulus similar to that observed in contractile protein and metabolic enzyme gene expression (see above). MyoD mRNA levels were unaffected in either muscle after 2 days of passive stretch (data not shown). In a previous study on the effect of stretch on qmfl, the avian homologue of MyoD, a marked elevation of this MRF was observed within hours, but levels had returned to control levels after 24 hours (Eppley, Kim & Russel, 1993). It is possible, therefore that, in our model, similar early responses occur in MyoD levels and this is currently under investigation.

In conclusion, both myogenin and MRF4 exhibited similar muscle specific responses, after disuse and passive stretch, to those we have previously observed in genes which contribute to the contractile properties of muscle. Our study suggests that the regulation of adult muscle phenotype by MRFs may not be confined to MyoD and myogenin as previously suggested, but MRF4 may also have a participatory role. Disuse causes dramatic muscle fibre atrophy which is prevented by passive stretch and over longer time periods causes fibre hypertrophy (Loughna, Goldspink & Goldspink, 1987). Whether MRFs also play a role in the determination of this muscle fibre size is at present unclear.

However, atrophy in fibres of transgenic animals has been observed where the inhibitory HLH protein Id, which forms heterodimers with MRFs and prevents DNA binding (Benezra *et al.*, 1990), is elevated, the degree of which relates directly to the level of expression of the transgene (Gunderson & Merlie, 1994). The potential of MRFs as regulatory factors in fibre size clearly requires further examination.

Conclusions

Mammalian muscles are composed of a heterogeneous mix of fibres with distinctly different phenotypes. The relative proportions of these sub-populations of fibres determine the contractile and metabolic properties of a particular muscle. We have examined the effects of immobilization in both the shortened (disuse) and lengthened (stretched) position upon levels of transcripts in slow and fast muscles. Results of initial studies using this model in which the levels of MyHC mRNAs were examined demonstrated that the response of individual transcripts to either disuse or passive stretch was dependent upon the muscle in which it was expressed. Expansion of this work to include transcripts of other contractile proteins and also metabolic enzymes demonstrates that they also exhibit a muscle-specific response to a particular stimulus.

The examination of a greater number of transcripts, however, permits a less reductionist view of the effects of disuse and passive stretch upon muscle phenotype. The plantaris, which is a fast muscle and is normally only active intermittently, is very little affected by disuse. It has been shown previously that inactivity produces considerably less atrophy in this muscle than slow muscles such as the soleus (Loughna *et al.*, 1986) and the present data demonstrate that the levels of both metabolic and contractile protein transcripts (predominantly fast) are similarly unaffected. In contrast the soleus muscle which has high levels of 'slow type' proteins and is a highly active postural muscle exhibits considerable change in phenotype even over this short time period. The decrease in levels of certain 'slow type' transcripts and the massive up-regulation of type 2B MyHC mRNA in response to disuse suggest that this muscle is moving to a faster phenotype. It thus appears that the slow phenotype of muscles such as the soleus must be maintained by high levels of activity which when removed cause the muscle to move to a faster phenotype.

It has been shown previously that passive stretch prevents the atrophy associated with disuse (Loughna *et al.*, 1986) and in the present studies we demonstrate that it greatly reduces the change towards a faster phenotype in the soleus. In the plantaris its effects are more marked where disuse had little or no effect upon phenotype, in this muscle stretch it produces a transition towards a slower phenotype by the down-regulation of fast transcripts. This stretch model is not, however, a very potent enhancer of muscle tension levels particularly for the soleus and its action reduces rapidly with time. A more potent activator of muscle tension, tenotomy of a synergistic muscle, has been shown to have more dramatic effects upon MyHC mRNA levels in both muscles (Morgan & Loughna, 1989). It is likely that such a model would produce a greater effect upon the levels of other muscle transcripts, enhancing transition to a slower phenotype more than the present stretch model.

This work further emphasizes the remarkable plasticity of adult skeletal muscle, but the mechanisms by which the phenotype of a muscle fibre is regulated are unclear. It has been suggested that MRFs may play a role in regulating adult muscle phenotype with myogenin being more prevalent in slow fibres (Hughes *et al.*, 1993). Consistent with this is the up-regulation of myogenin in the plantaris in response to passive stretch. More surprising, however, is the sensitivity of MRF4 to disuse and passive stretch in the soleus.

References

Alway, S.E., Winchester, P.K., Davis, M.E. & Gonyea, W.J. (1989). Regionalised adaptions and muscle fibre proliferation in stretch-induced enlargement. *Journal of Applied Physiology*, **66**, 771–81.

Benezra, R., Davis, R.L., Lockson, D., Turner, D.L. & Weintraub, H. (1990). The protein Id: a negative regulator of helix–loop–helix DNA binding proteins. *Cell*, **61**, 49–59.

Block, N.E. & Miller, J.B. (1992). Expression of MRF4, a myogenic helix–loop–helix protein, produces multiple changes in the myogenic program of BC3H-1 cells. *Molecular Cellular Biology*, **12**, 2484–92.

Brownson, C., Isenberg, H., Brown, W., Salmons, S. & Edwards, Y. (1988). Changes in skeletal muscle gene transcription induced by chronic stimulation. *Muscle and Nerve*, **11**, 1183–9.

Brownson, C., Little, P., Jarvis, J.C. & Salmons, S. (1992). Reciprocal changes in myosin isoform mRNAs of rabbit skeletal muscle in response to the initiation and cessation of chronic electrical stimulation. *Muscle and Nerve*, **15**, 694–700.

Davies, P.F. (1995). Flow-mediated endothelial mechanotransduction. *Physiological Reviews*, **75**, 519–60.

Davis, R.L., Cheng, P.F., Lassar, A.B. & Weintraub, H. (1990). The MyoD DNA binding domain contains a recognition code for muscle specific gene activation. *Cell*, **60**, 733–46.

Eppley, Z.A., Kim, J. & Russell, B. (1993). A myogenic regulatory gene, *qmfl*, is expressed by adult myonuclei after injury. *American Journal of Physiology*, **265**, C397–405.

Gunderson, K. & Merlie, J.P. (1994). Id-1 as a possible transcriptional mediator of muscle disuse atrophy. *Proceedings of the National Academy of Sciences, USA*, **91**, 3647–51.

Hughes, S.M., Taylor, J.M., Tapscott, S.J., Gurley, C.M., Carter, W.J. & Peterson, C.A. (1993). Selective accumulation of MyoD and myogenin mRNAs in fast and slow adult skeletal muscle is controlled by innervation and hormones. *Development*, **118**, 1137–47.

Izumo, S., Lompre, A.M., Matsouka, R., Koren, G., Schwartz, K. *et al.* (1987). Myosin heavy chain messenger RNA and protein isoform transitions during cardiac hypertrophy. *Journal of Clinical Investigation*, **79**, 970–7.

Izumo, S., Nadal-Ginard, B. & Mahadavi, V. (1986). All members of the myosin heavy chain multi-gene family respond to thyroid hormone in a highly tissue specific manner. *Science*, **231**, 597–600.

Izumo, S., Nadal-Ginard, B. & Mahadavi, V. (1988). Proto-oncogene induction and reprogramming of cardiac gene expression produced by pressure overload. *Proceedings of the National Academy of Sciences, USA*, **85**, 339–43.

Komuro, I. & Yazaki, U. (1993). Control of cardiac gene expression by mechanical stress. *Annual Reviews in Physiology*, **55**, 55–75.

Loughna, P.T. & Bates, P.C. (1994). Interactions between growth hormone and nutrition in hypophysectomized rats: skeletal muscle myosin heavy chain mRNA levels. *Biochemical and Biophysical Research Communications*, **198**, 97–102.

Loughna, P.T., Goldspink, G & Goldspink, D.F. (1986). Effects of inactivity and passive stretch on protein turnover in phasic and postural rat muscles. *Journal of Applied Physiology*, **61**, 173–9.

Loughna, P.T., Goldspink, D.F. & Goldspink, G. (1987). Effects of hypokinesia and hypodynamia upon protein turnover in hindlimb muscles of the rat. *Aviation, Space and Environmental Medicine*, **58**, A133–8.

Loughna, P.T., Izumo, S., Goldspink, G. & Nadal-Ginard, B. (1990). Disuse and passive stretch cause rapid alterations in expression of developmental and adult contractile protein genes in skeletal muscle. *Development*, **109**, 217–23.

McCormick, K.M. & Schultz, E. (1994). Role of satellite cells in altering myosin expression during skeletal muscle hypertrophy. *Developmental Dynamics*, **99**, 52–63.

Mahadavi, V., Strehler, E.E., Periasamy, M., Wieczorek, D.F., Izumo, S. & Nadal-Ginard, B. (1986). Sarcomeric myosin heavy chain gene family: organisation and pattern of expression. *Medical Science Sports Exercise*, **18**, 299–308.

Maltin, C.A., Delday, M.I., Campbell, G.P. & Hesketh, J.E. (1993). Clenbuterol mimics effects of innervation of myogenic regulatory factor expression. *American Journal of Physiology*, **265**, E176–8.

Morgan, H.E. & Baker K.M. (1991). Cardiac hypertrophy: mechanical, neural, and endocrine dependence. *Circulation*, **83**, 13–25.

Morgan, M.J. & Loughna, P.T. (1989). Work overload induced changes in fast and slow skeletal muscle myosin heavy chain gene expression. *FEBS Letters*, **255**, 427–30.

Olsen, E.N. (1990). MyoD family: a paradigm for development? *Genes Development*, **4**, 1454–61.

Parker, T.G. & Schneider, M.D. (1991). Growth factors, proto-oncogenes and plasticity of the cardiac phenotype. *Annual Reviews in Physiology*, **53**, 179–200.

Pette, D. & Dusterhoft, S. (1992). Altered gene expression in fast-twitch muscle induced by chronic low-frequency stimulation. *American Journal of Physiology*, **262**, R333–8.

Schwartz, K., de la Bostie, D., Bouvert, P., Oliviero, P., Alonso, S. *et al.* (1986). α skeletal muscle action mRNAs accumulate in hypertrophied adult rat hearts. *Circulation Research*, **59**, 551–5.

Sola, O.M., Christensen, D.L. & Martin, A.W. (1973). Hypertrophy and hyperplasia of adult chicken anterior latissimus dorsi muscles following stretch with and without denervation. *Experimental Neurology*, **41**, 76–100.

Swyngehdauw, B. (1986). Developmental and functional adaption of contractile proteins in cardiac and skeletal muscles. *Physiological Reviews*, **66**, 710–71.

Tapscott, S.J., David, R.L., Tayler, M.J., Cheng, P., Weintraub, H. & Lassar, A.B. (1988). MyoD 1: a nuclear phosphoprotein requiring a *myc* homology region to convert fibroblasts to myoblasts. *Science*, **242**, 405–11.

Watson, P.A. (1991). Function follows form: generation of intracellular signals by cell deformation. *FASEB Journal*, **5**, 2013–19.

Winchester, P.K. & Gonyea, W.J. (1992). Regional injury and the terminal differentiation of satellite cells in stretched avian slow muscle. *Developmental Biology*, **151**, 459–72.

N.C. STICKLAND and C.M. DWYER

The pre-natal influence on post-natal muscle growth

Summary

Animals with more muscle fibres grow faster and more efficiently than animals with fewer fibres. In mammals, the number of muscle fibres in a given muscle is determined pre-natally. In other words, post-natal growth potential is determined *in utero*.

We have studied the pre-natal determination of muscle fibre number using nutritional manipulation in pregnant animals. Nutritional restriction throughout gestation in pregnant guinea pigs causes a 20% reduction in fetal muscle fibre number by birth. Maternal and fetal serum levels of IGF-1 are also reduced as is fetal IGF-2. Cortisol is inversely related to IGF-1 levels, but thyroid hormones do not seem to be important until at least late gestation. Maternal undernutrition impairs the expansion of the peripheral labyrinth (exchange surface) of the placenta causing a reduction in placental efficiency in early gestation. More recent work has demonstrated that nutritional restriction in the early stages of gestation only (and then *ad lib* to birth) is sufficient to reduce muscle fibre number in the offspring. Allied work in the pig has also demonstrated a very early critical stage in gestation when muscle fibre number can be permanently affected as well as IGF-1 levels at birth.

It is suggested that nutritional status very early in gestation can affect post-natal growth performance and that this is mediated through maternal IGFs affecting placental growth. This in turn affects fetal IGFs which influence developing muscle tissue.

Introduction

Muscle mass is determined by muscle fibre number and muscle fibre size. Various factors, including exercise and nutrition can affect the size and types of fibres in post-natal animals, but fibre number is unaffected (Stickland, Widdowson & Goldspink, 1975; Goldspink &

Ward, 1979). In the pig, muscle fibre hyperplasia is completed during gestation and is fixed from the time of birth (Staun, 1963; Stickland & Goldspink, 1973). However, fibre number can be affected by conditions *in utero* including maternal nutrition (Wigmore & Stickland, 1983) and innervation (McLennan, 1983). As muscle fibre number is probably the most important determinant of muscle mass (Miller, Garwood & Judge, 1975) it is evident that pre-natal conditions which affect fibre number determinations may have a significant effect on post-natal muscle growth. Although selection for muscle fibre size in the pig can also correlate with an increase in muscle size after two generations (Fiedler, 1988) there is a corresponding decrease in meat quality. Ashmore (1974) showed that the correlation between large fibre size and poor meat quality appeared to be due to a relative increase in fast fibres with glycolytic rather than oxidative metabolism. It would seem, on balance, that high muscle fibre number (which correlates with smaller fibres) is the most relevant parameter in relation to improved muscle growth and quality. This review highlights the importance of muscle fibre number and also explains how this parameter may be affected *in utero* and the consequences of this on post-natal growth. Studies on post-natal growth were largely carried out on the pig. However, mechanisms have been investigated mainly on laboratory animals, particularly the guinea pig.

Muscle fibre number and post-natal growth

Fast-growing strains of pigs, mice and quail have more muscle fibres than slower-growing strains (Ezekwe & Martin, 1975; Miller *et al.*, 1975; Fowler *et al.*, 1980). This was also found to be true when Dwyer *et al.* (1993) considered pigs within the same strain. Sixty-six Large White × Landrace pigs from seven litters were reared under similar commercial conditions and weighed at monthly intervals from birth to slaughter (at about 80 kg). At slaughter the semitendinosus muscle was removed from each pig and the total number of muscle fibres was estimated. It was found that average daily gain (ADG) was significantly correlated with birth weight but not with fibre number for the initial growth period up to 25 kg body weight. However, for growth from 25 to 80 kg only fibre number was correlated with ADG. Furthermore, during this period, there was also a significant correlation between fibre number and gain/feed ratio. A relationship between birth weight and growth rate to weaning has been shown in other studies (Campbell & Dunkin, 1982). It is probable that small pigs compete less effectively for nutrition. Growth rates can, in fact, be improved by separating piglets into similar weight groups (England, 1974) which suggests that the correlation between birth weight and early growth may be related to feed intake. The later relationship after 25 kg between fibre number (not birth weight) and growth rate supports the notion that growth at this stage is more determined by the pig's genotype (Blunn, Baker & Hanson, 1953). It is known that pigs selected for fast growth are more efficient in feed conversion and contain less fat than unselected pigs (Campbell & Taverner, 1988). Pigs with more muscle fibres also exhibit less fat (Stickland & Goldspink, 1975) and this is consistent with more efficient growth. Conversely obese pigs exhibit slow growth and have low fibre numbers (Hausman, Campion & Thomas, 1983). At equivalent live weights pigs with more fibres also have thinner fibres (Fiedler, 1988); delivery of nutrients to thinner fibres may be more efficient due to smaller diffusion distances. Taken overall, it seems therefore that animals with high fibre number grow faster and more efficiently than those with a low fibre number.

In order to understand the factors which might affect fibre number determination it is necessary to understand the process of myogenesis in terms of primary and secondary fibre formation.

Myogenesis: primary and secondary myofibres

Muscle fibres form by fusion of mononucleate myoblasts into multinucleated myofibres. Myoblasts proliferate in presumptive muscle regions and there is some evidence, in avian species at least (Stockdale & Miller, 1987), that there are two populations of myogenic precursor cells which develop into different fibre types. Pre-natally muscle fibres develop as two distinct populations. Fibres which form first are primary myofibres, which form by a nearly synchronous fusion of myoblasts (Harris *et al.*, 1989). These primary fibres provide a framework for the formation of the larger population of smaller secondary myofibres which form by asynchronous fusion of myoblasts over a longer time period. During their early formation, secondary fibres have very close mechanical contact with primary fibres (Duxson & Usson, 1989). In most porcine muscles the two populations can be distinguished by adenosine triphosphatase activity with primaries reacting as slow fibres and secondaries as fast (Fig. 1.). In the pig, only primary fibres are present at 40 days gestation and their number increases to a maximum at about 60 days. Secondary myofibre formation takes place from about 54 up to 90 days gestation. In the semitendinosus muscle about 25 secondaries form around each primary myofibre (Wigmore & Stickland, 1983). During later pre-natal development, and also during

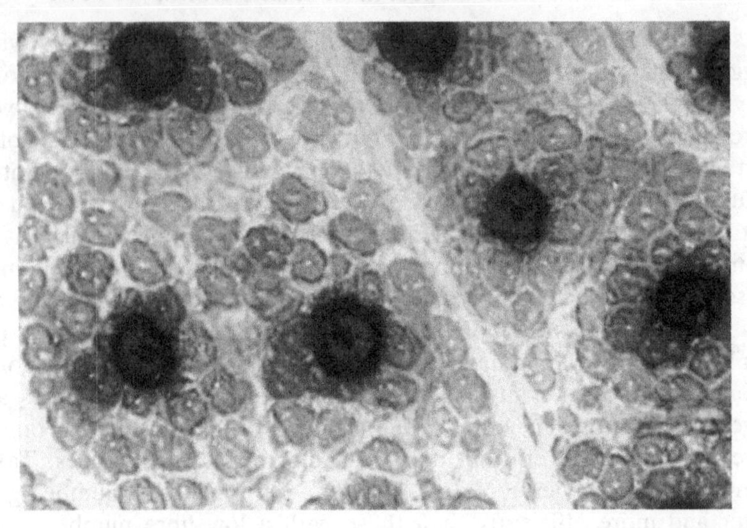

Fig. 1. Transverse section of porcine semitendinosus muscle reacted for acid-stable adenosine triphosphatase activity. The large positively reacted fibres (slow) are the primary myofibres and the remainder are secondary myofibres.

post-natal growth, some of the secondaries differentiate into slow fibres so that clusters of slow fibres (one of which was a primary myofibre) are surrounded by fast fibres. This arrangement of fibre bundles (sometimes called metabolic bundles) persists in the postnatal pig so that, even in mature pigs, it is possible to determine the numbers of primaries and secondaries which contributed to the development of a given muscle.

Sources of variation in muscle fibre number

In a survey of 48 Large White pigs from five litters (Dwyer & Stickland, 1991), it was found that primary fibre number in *m.semitendinosus* was responsible for the variation in total muscle fibre number between litters. Within litters, both primary number and secondary:primary ratio contributed to differences in fibre number. However, when only the largest and smallest extremes of the litters were compared, variation in fibre number was due only to secondary:primary ratio differences; this confirms the results of Handel & Stickland (1987) who also showed that smallest littermates contained relatively more slow muscle fibres than larger littermates. The smallest pigs in a newborn litter are almost certainly a consequence of undernutrition *in utero*. This is highlighted by the U-shaped distribution of fetal weights in each uterine horn which is believed to be a consequence of differing nutrition (McLaren & Mitchie, 1960; Perry & Rowell, 1969). The development of muscle fibre number was studied in largest and smallest (excluding the runt) littermates during gestation by Wigmore & Stickland (1983). No difference was found in the number of primary fibres but a difference was observed in the secondary fibre population from about 65 days onwards which produced a difference of 17% in the numbers by birth. At the time of rapid secondary fibre hyperplasia primary fibres in the smaller fetuses were smaller in diameter. It was speculated that the small size may restrict the available surface area for secondary fibre formation. However, a difference in myoblast proliferation must also be a contributory factor as the muscles exhibit a difference in DNA content (Wigmore & Stickland, 1984). In the mouse also, Penney *et al.* (1983) concluded that QL mice (selected for large body weight at six weeks of age) had more muscle fibres than QS mice (selected for small body weight) due to a difference in nuclear division.

Taken as a whole, the results suggest that primary fibre number is a relatively more fixed genetic component than secondary fibre number and is therefore more indicative of the genotype of an animal. This is confirmed in a study by Stickland & Handel (1986) which showed that primary fibre number was the major contributor to the muscle fibre number difference seen between large and small strains of pigs. Secondary fibre number would appear to be more vulnerable to environmental factors *in utero*, including nutrition. This is confirmed by various experiments including those by Ward & Stickland (1991) which showed that undernutrition of pregnant guinea pigs to 60% of *ad lib* throughout gestation produced a 26% deficit in secondary fibre numbers, but no affect on primary numbers, in biceps brachii muscles of the neonatal offspring (Fig. 2.). Interestingly, the soleus (a slow twitch muscle) did not suffer this deficit in fibre number. Dwyer & Stickland (1992*b*) concluded that the soleus muscle was less susceptible to prenatal undernutrition than the putative fast twitch biceps brachii due to its higher proportion of primary fibres. Primary fibres are unaffected by undernutrition; this may be due to their lower and earlier rate of formation compared to secondary fibres.

Possible mechanisms whereby nutrition affects fibre number

The mechanism whereby undernutrition affects fibre number determination has been investigated in the guinea pig. In pregnant guinea pigs

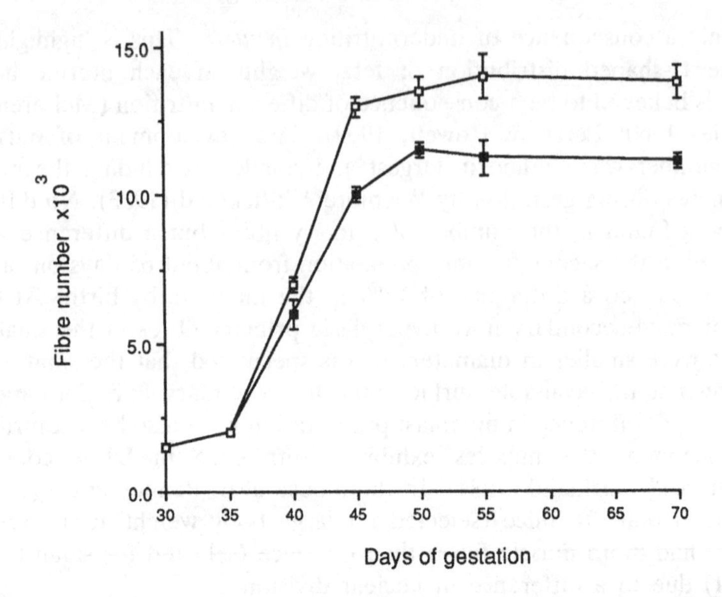

Fig. 2. Change in biceps brachii fibre number with gestational age in fetuses from *ad lib* fed mothers (-□-) and from 60% of *ad lib* fed mothers (-■-). Error bars, ± SEM.

restricted to 60% of an *ad lib* diet throughout gestation placental weight and the exchange surface area (peripheral labyrinth) was reduced by 38% and 33% respectively (Fig. 3 and Dwyer *et al.*, 1992). Maternal undernutrition results in an impaired or delayed expansion of the peripheral labyrinth causing a reduction in placental efficiency in early gestation (Fig. 3). Towards the end of gestation the weight of the peripheral labyrinth is significantly correlated with fetal weight suggesting that it is the size of the peripheral labyrinth which limits fetal size. In another experiment (Dwyer & Stickland, 1992*a*) undernutrition was found to reduce maternal and fetal serum levels of IGF-1 as well as fetal levels of IGF-2 (Fig. 4). Cortisol appeared to be inversely related to IGF-1 levels but thyroid hormones do not seem to be important until at least late gestation and therefore unlikely to affect muscle fibre number. The IGFs may be important regulators of muscle fibre number as they have been shown to stimulate myoblast proliferation *in vitro* (Ewton & Florini, 1980). It seems likely therefore that nutritional affects on muscle fibre number determination are mediated through effects on fetal levels of IGFs.

In a recent experiment by Dwyer *et al.* (1995) it was shown that a 40% nutritional restriction to pregnant guinea pigs from conception to

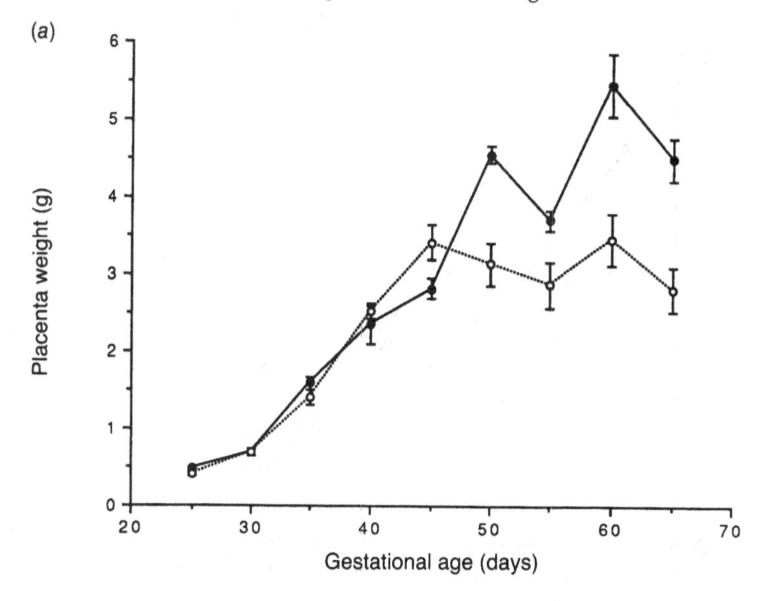

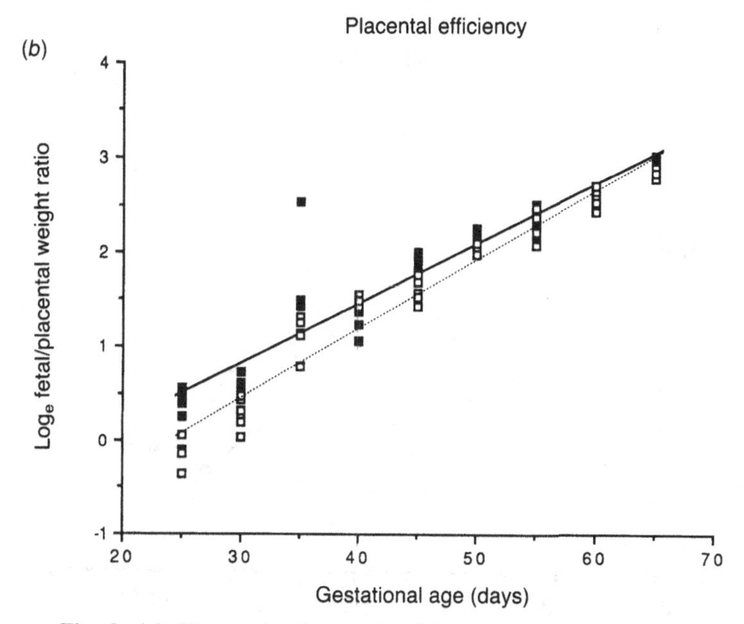

Fig. 3. (*a*) Change in placental weight of control (–●–) and restricted (··○··) guinea pig fetuses. Values are litter means (± SEM). (*b*) Regression of log$_e$ feto-placental weight ratio against gestational age for control (–■–) and restricted (··□··) fetuses.

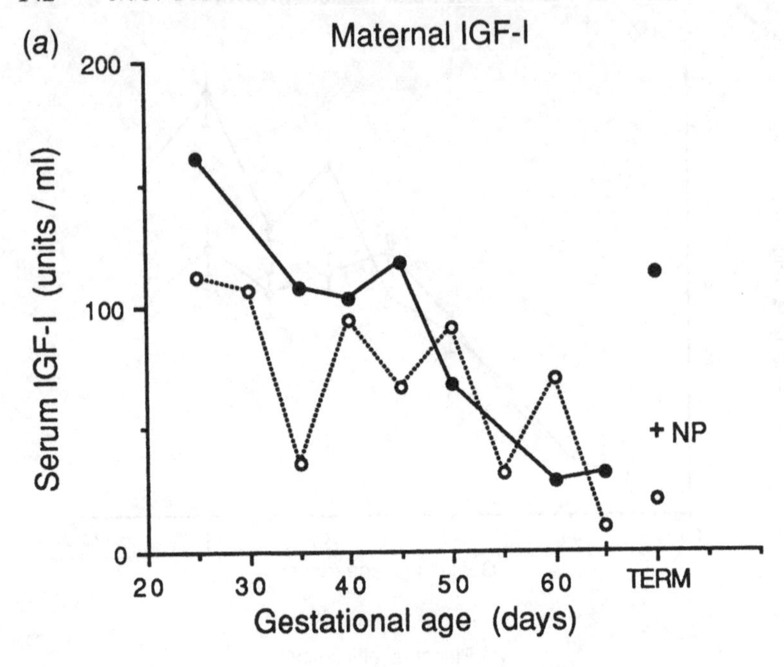

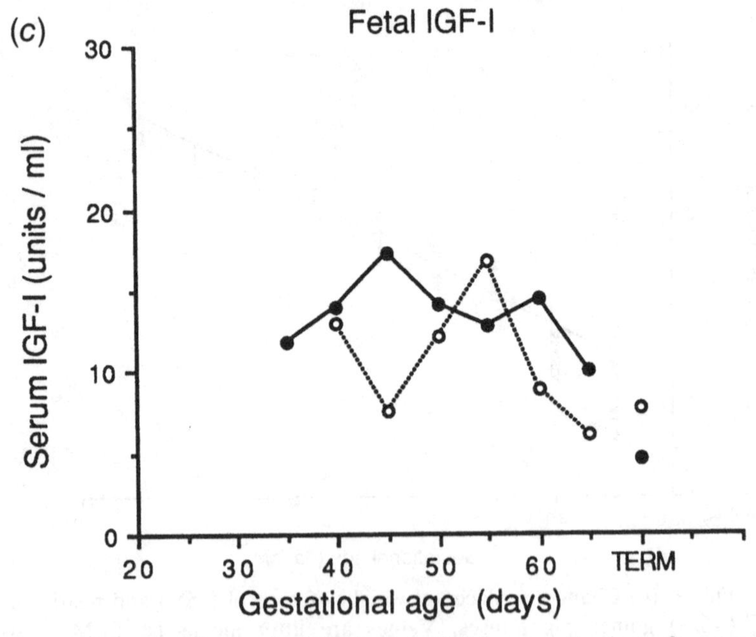

Fig. 4. Change in IGF-1 and IGF-2 concentration of maternal and fetal sera with gestational age, for control (–●–) and restricted (··○··) samples. NP = non-pregnant female concentration.

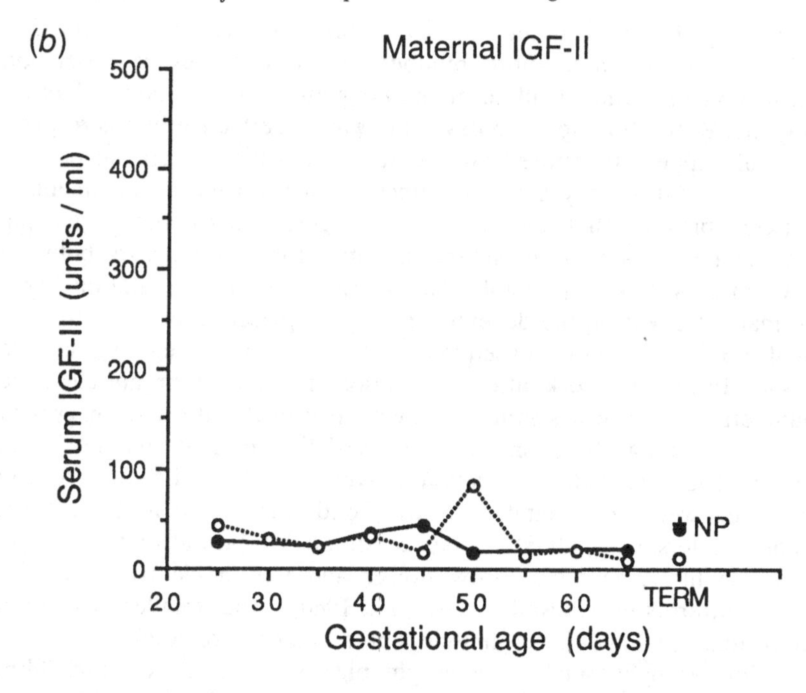
(b)
Maternal IGF-II
Serum IGF-II (units / ml)
500
400
300
200
100
0
20
30
40
50
60
TERM
NP
Gestational age (days)

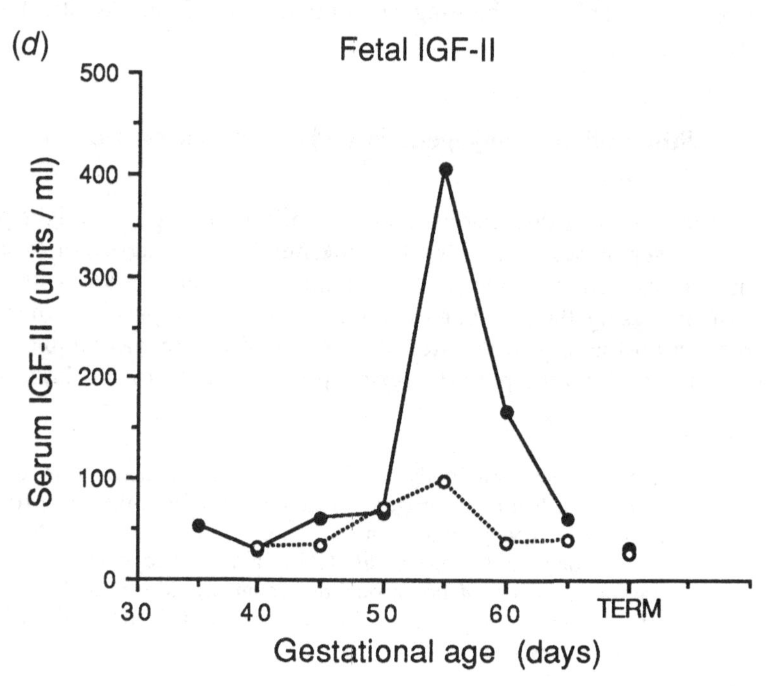
(d)
Fetal IGF-II
Serum IGF-II (units / ml)
500
400
300
200
100
0
30
40
50
60
TERM
Gestational age (days)

25 days gestation (and then *ad lib* to term) produced the same 20–28% deficit in muscle fibre number in the neonates as restriction throughout gestation (full term in the guinea pig is about 70 days) (Fig. 5). Restriction up to 15 days only was not effective in this respect. It would appear therefore that inadequate nutrition at critical periods in early gestation may have a significant effect on the development of muscle fibres which takes place later in gestation (in the guinea pig the maximum rate of secondary myofibre formation occurs between days 35 to 45). It is probable that restriction in early gestation has a permanent effect on the development of the placenta thereby affecting fetal IGF levels with consequent effects on the developing muscle tissue. From the work already mentioned a deficit in muscle fibre numbers may have a significant effect on growth rate and on growth efficiency. It has been shown that restriction of maternal intake in pregnant sows leads to a reduction in average daily gain of the progeny from ten weeks post-natal onwards (Pond, Mersmann & Yen, 1985; Pond & Mersmann, 1988). Under conditions of natural undernutrition within a litter, runt pigs grow slower and less efficiently than their larger littermates (Powell & Aberle, 1980). The smaller pigs in a litter tend to have fewer muscle fibres (Handel & Stickland, 1987) but, interestingly, small birth weight pigs with a relatively high fibre number are capable of exhibiting catch-up growth (Handel & Stickland, 1988).

Stimulation of myogenesis and effects on post-natal growth

From the foregoing discussions it would appear that lighter weight pigs contain fewer muscle fibres due to undernutrition *in utero*. An experiment was carried out (Dwyer, Stickland & Fletcher, 1994) with the aim of increasing the number of fibres in the lighter pigs and thereby increasing postnatal growth rate. Maternal feed intake was doubled for one of three different periods during pregnancy: 25 to 50 days (HE

Fig. 5. (*a*) Neonatal body weight and (*b*) Fibre number in biceps brachii at birth for the progeny of guinea pigs fed either a restricted diet up to 15 days gestation and then *ad lib* (VER), a restricted diet up to 25 days and then *ad lib* (ER), a restricted diet throughout gestation (TR), or *ad lib* throughout (control). Means and standard errors are given. Significant differences ($P<0.05$) are indicated by differing letters.

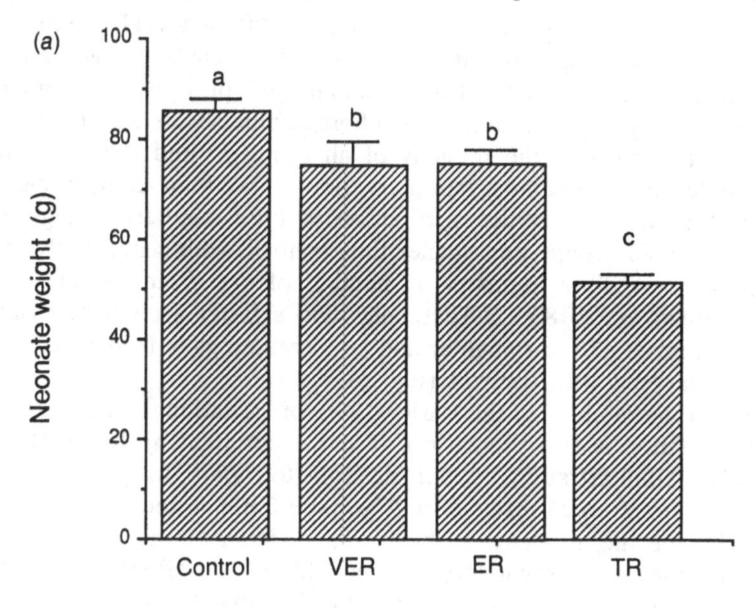
(a)
100
80
60
40
20
0
Neonate weight (g)
a
b
b
c
Control
VER
ER
TR

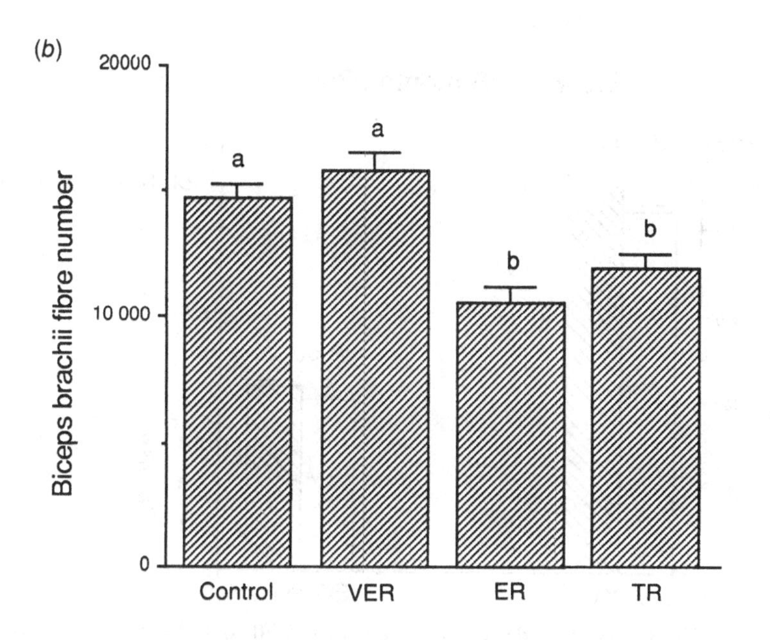
(b)
20000
10 000
0
Biceps brachii fibre number
a
a
b
b
Control
VER
ER
TR

group), 50 to 80 days (HL group) or 25 to 80 days (HT group). The offspring were slaughtered at 5 weeks and their semitendinosus muscles sectioned and analysed. The variability of primary fibre number between the groups meant that total fibre numbers were not significantly different. However, the progeny of all supplemented sows contained a greater mean secondary:primary fibre ratio than control pigs. It was also evident that the distribution of fibre numbers in pigs from supplemented groups was smaller than controls with significantly fewer pigs with low fibre numbers. A number of pigs from the HT group were grown on to 80 kg. These pigs had a faster growth rate (about 10%) from day 70 to slaughter and an increased gain:feed ratio (about 8%) compared with controls (Fig. 6).

The vulnerability of the early stages of gestation to fibre number determination is also shown in a recent experiment by Rehfeldt *et al.* (1993). Administration of porcine somatotrophin to pregnant sows during days 10 to 24 resulted in 27% more fibres in the semitendinosus muscles of the progeny at birth. This increase was due solely to an increase in the secondary : primary fibre ratio (Rehfeldt, personal communication). Also of interest is the fact that the hormone treated

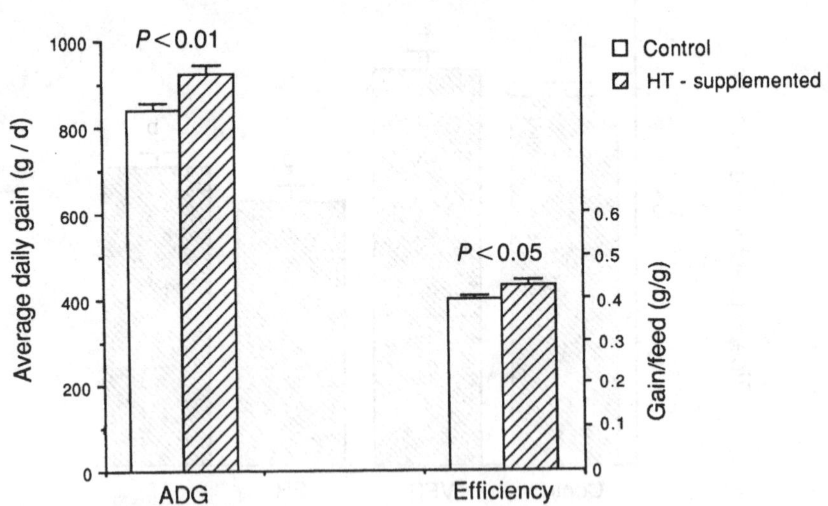

Fig. 6. Average daily gain (ADG) from 70 to 130 days and gain/feed ratio for progeny of nutritionally supplemented sows (HT) compared with controls. Means and standard errors are given. Significances of differences are indicated.

sows produced piglets with significantly increased levels of serum IGF-1 at birth. This may indicate faster postnatal growth as Buonomo *et al.* (1987) found that fast-growing strains of pigs contained higher serum IGF-1 levels than slow-growing pigs.

Conclusions

This chapter has highlighted the importance of muscle fibre number to postnatal growth rate and feed conversion efficiency. It is clear that levels of maternal nutrition, particularly in early gestation, may have a very significant effect on muscle fibre number development *in utero*. Sows given extra feed during early gestation produce pigs with improved growth rate and greater feed conversion efficiency. The IGFs have been implicated in regulating fetal and, in particular placental growth, and may be influential in muscle fibre number determination and potential for post-natal growth.

References

Ashmore, C.R. (1974). Phenotypic expression of muscle fibre types and some implications to meat quality. *Journal of Animal Science*, **38**, 1158–64.

Blunn, C.T., Baker, G.N. & Hanson, L.E. (1953). Heritability of gain in different growth periods in swine. *Journal of Animal Science*, **12**, 39–50.

Buonomo, F.C., Lanterio, T.J., Baile, C.A. & Campion, D.R. (1987). Determination of insulin-like growth factor-1 (IGF-1) and IGF binding protein levels in swine. *Domestic Animal Endocrinology*, **4**, 23–31.

Campbell, R.G. & Dunkin, A.C. (1982). The effects of birth weight and level of feeding in early life on growth and development of muscle. *Animal Production*, **35**, 185–92.

Campbell, R.G. & Taverner, M.R. (1988). Genotype and sex effects on the relationship between energy intake and protein deposition in growing pigs. *Journal of Animal Science*, **66**, 676–86.

Duxson, M.J. & Usson, Y. (1989). Cellular insertion of primary and secondary myotubes in embryonic rat muscle. *Development*, **107**, 243–51.

Dwyer, C.M., Fletcher, J.M. & Stickland, N.C. (1993). Muscle cellularity and post-natal growth in the pig. *Journal of Animal Science*, **71**, 3339–43.

Dwyer, C.M., Madgwick, A.J.A., Crook, A.R. & Stickland (1992). The effect of maternal undernutrition on the growth and development of the guinea pig placenta. *Journal of Developmental Physiology*, **18**, 195–302.

Dwyer, C.M., Madgwick, A.J.A., Ward, S.S. & Stickland, N.C. (1995). The effect of maternal undernutrition in early gestation on the development of fetal myofibres in the guinea pig. *Reproduction, Fertility and Development*, **7**, 1285–92.

Dwyer, C.M. & Stickland, N.C. (1991). Sources of variation in myofibre number within and between litters of pigs. *Animal Production*, **52**, 527–33.

Dwyer, C.M. & Stickland, N.C. (1992*a*). The effects of maternal undernutrition on maternal and fetal serum insulin-like growth factors, thyroid hormones and cortisol in the guinea pig. *Journal of Developmental Physiology*, **18**, 303–13.

Dwyer, C.M. & Stickland, N.C. (1992*b*). Does the anatomical location of a muscle affect the influence of undernutrition on muscle fibre number? *Journal of Anatomy*, **181**, 373–6.

Dwyer, C.M., Stickland, N.C. & Fletcher, J.M. (1994). The influence of maternal nutrition on muscle fibre number development in the porcine fetus and on subsequent post-natal growth. *Journal of Animal Science*, **72**, 911–17.

England, D.C. (1974). Husbandry components in prenatal and perinatal development in swine. *Journal of Animal Science*, **38**, 1045–9.

Ewton, D.Z. & Florini, J.R. (1980). Relative effects of the somatomedins, MSA and growth hormone on myoblasts and myotubes in culture. *Endocrinology*, **106**, 577–83.

Ezekwe, M.O. & Martin, R.J. (1975). Cellular characteristics of skeletal muscles in selected strains of pigs and mice and the unselected controls. *Growth*, **39**, 95–106.

Fiedler, I. (1988). Leistungsfruherkerming von Fleischatstz und Fleischbeschaffenheit durch Muskelfasermerkmate. *Tag-ber. Akad. Landwirtsch. Wiss. Berlin*, **268**, 187–96.

Fowler, S.P., Campion, D.R., Marks, H.L. & Reagan, J.O. (1980). Analysis of skeletal muscle response to selection for rapid growth in Japanese Quail. *Growth*, **44**, 235–52.

Goldspink, G. & Ward, P. (1979). Changes in rodent fibre types during post-natal growth, undernutrition and exercise. *Journal of Physiology*, **296**, 453–69.

Handel, S.E. & Stickland, N.C. (1987). Muscle cellularity and birth weight. *Animal Production*, **44**, 311–17.

Handel, S.E. & Stickland, N.C. (1988). Catch-up growing in pigs: a relationship with muscle cellularity. *Animal Production*, **47**, 291–5.

Harris, A.J., Duxson, M.J., Fitzsimmons, R.B. & Rieger, F. (1989). Myonuclear birthdates distinguish the origins of primary and secondary myotubes in embryonic skeletal muscles. *Development*, **107**, 771–84.

Hausman, G.J., Campion, D.R. & Thomas, G.B. (1983). Semitendinosus muscle development in several strains of fetal and perinatal pigs. *Journal of Animal Science*, **57**, 1608–17.

McLaren, A. & Mitchie, D. (1960). Control of prenatal growth in mammals. *Nature*, **187**, 363–5.

McLennan, I.S. (1983). Neural dependence and independence of myotube production in chick hindlimb muscles. *Developmental Biology*, **98**, 287–94.

Miller, L.R., Garwood, V.A. & Judge, M.D. (1975). Factors affecting porcine muscle fiber type, diameter and number. *Journal of Animal Science*, **41**, 66–7.

Penney, P.K., Prentis, P.F., Marshall, P.A. & Goldspink, G. (1983). Differentiation of muscle and the determination of ultimate tissue size. *Cell and Tissue Research*, **228**, 365–88.

Perry, J.S. & Rowell, J.G. (1969). Variations in foetal weight and vascular supply along the uterine horn of the pig. *Journal of Reproduction and Fertility*, **19**, 527–34.

Pond, W.G. & Mersmann, H.J. (1988). Comparative responses of lean or genetically obese swine and their progeny to severe feed restriction during gestation. *Journal of Nutrition*, **118**, 1223–34.

Pond, W.G., Mersmann, H.J. & Yen, J-T. (1985). Severe food restriction of pregnant swine and rats: effect on postweaning growth and body composition of progeny. *Journal of Nutrition*, **115**, 179–89.

Powell, S.E. & Aberle, E.D. (1980). Effects of birth weight on growth and carcass composition of swine. *Journal of Animal Science*, **50**, 860–8.

Rehfeldt, Ch., Fiedler, I., Weikard, R., Kanitz, E. & Ender, K. (1993). It is possible to increase skeletal muscle fibre number *in utero*. *Bioscience Reports*, **13**, 213–20.

Staun, H. (1963). Various factors affecting number and size of muscle fibres in the pig. *Acta Agriculturae Scandinavica*, **13**, 293–322.

Stickland, N.C. & Goldspink, G. (1973). A possible indicator muscle for the fibre content and growth characteristics of porcine muscle. *Animal Production*, **16**, 135–46.

Stickland, N.C. & Goldspink, G. (1975). A note on porcine skeletal muscle parameters and their possible use in early progeny testing. *Animal Production*, **21**, 93–6.

Stickland, N.C. & Handel, S.E. (1986). The numbers and types of muscle fibres in large and small breeds of pigs. *Journal of Anatomy*, **147**, 181–9.

Stickland, N.C., Widdowson, E.M. & Goldspink, G. (1975). Effects of severe energy and protein deficiencies on the fibres and nuclei in skeletal muscles of pigs. *British Journal of Nutrition*, **34**, 421–8.

Stockdale, F.E. & Miller, J.B. (1987). The cellular basis of myosin heavy chain isoform expression during development of avian skeletal muscle. *Developmental Biology*, **123**, 1–19.

Ward, S.S. & Stickland, N.C. (1991). Why are slow and fast muscles differentially affected during prenatal undernutrition? *Muscle and Nerve*, **14**, 259–67.

Wigmore, P.M.C. & Stickland, N.C.A. (1983). Muscle development in large and small pig fetuses. *Journal of Anatomy*, 235–45.
Wigmore, P.M.C. & Stickland, N.C. (1984). DNA, RNA and protein in skeletal muscle of large and small pig fetuses. *Growth*, **47**, 67–76.

GUDRUN E. MOORE, PHILLIP R. BENNETT, ZEHRA ALI, REHAN U. KHAN and JANET I. VAUGHAN

Genomic imprinting and intrauterine growth retardation

The role of genomic or gametic imprinting and its association with growth has become an exciting research area during the last 10 years. Studies in both man and mouse have implicated chromosomal regions and specific genes that are imprinted such that the expression of the same gene is different depending on its parental origin. When the normal imprinted pattern of genes in some of these regions is altered through deletions, uniparental disomy or localized mutations affecting specific gene expression, specific growth disorders are observed.

Intrauterine growth retardation

Intrauterine growth retardation (IUGR) is one of the three major causes of perinatal and childhood morbidity (Dobson, Abell & Beischer, 1981). Only prematurity and major malformations numerically outrank it. In many cases the aetiology is unknown, making management and prognosis difficult. IUGR can be divided broadly into three aetiological groups: cases associated with pre-eclampsia or vascular disorders; those found to have syndromes associated with chromosomal abnormalities; and a heterogeneous 'idiopathic' group.

The pattern of intrauterine growth that defines IUGR occurs when a fetus is observed to be small for gestational age (SGA) and growing slowly as defined by ultra-sound evidence. The post-natal characteristics have traditionally been a SGA baby with a fairly typical clinical appearance suggesting poor intrauterine nutrition. The prevalence of IUGR is difficult to assess and depends heavily on how the limits to the growth parameters are defined. It is defined generally by birthweight below the 10th percentile in accordance with the gestational age for infants born in the same community (Kitchen, 1968). However about 75% of these show catch-up growth during the first four years of life (Fancourt *et al.*, 1976).

Genomic imprinting

Definition of Genomic Imprinting

There is increasing evidence that the function of a chromosome region or specific gene may differ depending upon whether it is maternally or paternally derived. Originally described by Crouse in 1960, in the homopteran scale insect *Sciara*, this has been termed genomic or gametic imprinting (Surani, Barton & Norris, 1984; Barlow, 1994). Differences between the maternally and paternally derived chromosomes, known as 'imprinting effects', remain fixed through successive mitotic divisions. At meiosis, the chromosomes must be newly imprinted depending on the gender of original. The actual nature of the imprint is unknown but one of the suggested and experimentally studied mechanisms is that of DNA methylation. For example, mutant mice with impaired methylation have been reported. Moreover, the imprints of three known imprinted genes *Igf-2r, Igf-2* and *H19* were lost, resulting in the expression or repression of *Igf-2r, Igf-2 and H19* of both parental copies and death of the mutant embyros (Li, Beard & Jaenisch, 1993).

Observations in mice

In the mouse the development of pronuclear transplantation techniques has given rise to some convincing evidence for the role genomic imprinting plays in early development of the embryo and placenta. Pronuclei, maternal and paternal, can be identified, removed and then reintroduced back into the zygote (McGrath & Solter, 1983). Using this method the development of gynogenetic embryos, containing only maternal chromosome pairs, and androgenetic embryos, containing only paternal chromosome pairs, can be compared. Neither zygote was viable; however, there were developmental differences between the two. Gynogenetic development was almost completely embryonic while androgenetic development appeared almost exclusively extraembryonic. Therefore, in early mouse development selected paternally derived genes control the development of the placenta whilst other maternally derived genes play a more important role in the development of the embryo (McGrath & Solter, 1984; Surani *et al*., 1984). These observations suggest that the parental origin of key chromosomes or genes may play a role in fetal growth and development.

Observations in humans

One of the best examples of genomic imprinting in humans is complete hydatidiform mole (Lawler, 1984). This abnormal conceptus comprises entirely of hydropic trophoblast tissue with no fetal tissue. Its genotype is paternally derived (46XX), demonstrating that, although such a genotype is incompatible with life, it will produce extraembryonic tissue. Moreover, triploid conceptuses show imprinting effects. Triploids with two sets of paternally derived chromosomes usually have large overgrown placentas and fetal IUGR, whereas triploids with two sets of maternally derived chromosomes abort early in gestation with underdeveloped placentas (MacFadden & Kalousek, 1991). There are several examples of human deletion disorders where the phenotype depends upon the parental origin of the relevant chromosome. For example, deletion of 15q11–13 of the paternally derived chromosome causes Prader–Willi syndrome, but deletion of the same region of the maternal chromosome leads to Angelman syndrome (Hall, 1990). Prader–Willi syndrome is characterized by hypotonia in infancy, obesity, hypogonadotrophic hypogonadism, small hands and feet, mental retardation and specific facies whereas Angelman syndrome individuals are hyperactive, happy disposition with unusual laughter, unusual seizures, specific facies and repetitive symmetrical ataxic movements (Hall, 1992). Whether the infant has lost this region from the father or mother greatly changes the phenotype observed.

Uniparental disomy and IUGR

Mechanisms giving rise to uniparental disomy

Uniparental disomy (UPD) occurs when a pair of homologous chromosomes are both inherited from the same parent. There have been several mechanisms postulated. These include (i) fertilization of a nullisomic by a disomic gamete; (ii) chromosome duplication in a monosomic somatic cell after postzygotic loss of a homologous chromosome; (iii) loss of supernumerary chromosome from a trisomic cell leaving two homologues from the same gamete (termed 'trisomic rescue') (Engel & DeLozier-Blanchet, 1991). High rates of aneuploidy in gametes of up to 50% in oocytes and 5% in spermatozoa suggest that uniparental disomy may be fairly common although it is not commonly reported because UPD detection requires DNA analysis rather than cytogenetic analysis (Martin *et al.*, 1987; Wramsby, Fredga & Leidholm, 1987). If 'trisomic rescue' is the most common mechanism then it is more likely that maternal UPD should be observed.

Observations in mice

By crossing animals with various Robertsonian and reciprocal translocations, Cattanach and Beechey, (1990) selectively bred mice which have

UPD. After analysing the development of these mice which were disomic for specific regions delineated by the translocations, it was possible to construct a 'non-complementation' map revealing that some chromosomes or chromosome segments require both parental complements for viability. Although, in mice, UPD for certain chromosomes is lethal in other cases the mice are structurally normal but growth disordered. Mouse maternally derived UPD (mUPD) for chromosome 7 (proximal) causes placental growth retardation whilst paternally derived UPD (pUPD) produces normal animals. pUPD Ch 7 (distal) causes IUGR whilst mUPD chromosome 7 (distal) is lethal. In contrast, mUPD Ch 11 (proximal), causes IUGR whilst pUPD 11 (proximal) causes macrosomia (Cattanach & Kirk 1985; Cattanach & Beechey, 1990). Since these animals are otherwise normal, the mechanism must be growth specific and affect either placenta and fetus or just the placenta with secondary fetal effects.

Observations in humans

Detection of UPD requires analysis of DNA polymorphisms to trace the parental origin of each chromosome as cytogenetic studies will be normal. There are now many examples of UPD in humans (See Table 1). For example, as mentioned above, Prader–Willi syndrome is usually caused by a cytogenetically visible deletion detected in the paternally derived chromosome 15 (15q11–13), but 25% have been found to be

Table 1 *Examples of uniparental disomy found in man*

	Chromosome	Phenotype
5	pUPD	Spinal muscular atrophy (Brzustowicz *et al.*, 1994)
7	mUPD	Cystic fibrosis/IUGR (Spence *et al.*, 1988; Voss *et al.*, 1989)
11	pUPD	Beckwith–Wiedemann syndrome Wilm's tumour (Henry *et al.*, 1991)
14	mUPD	Short stature and developmental delay (Antonarakis *et al.*, 1993)
15	pUPD	Prader–Willi syndrome
	mUPD	Angelman syndrome (Mascari *et al.*, 1992; Nicholls, 1994)
16	mUPD	IUGR (Bennett *et al.*, 1992)
21	mUPD	Early embryonic failure (Henderson *et al.*, 1994)
X	pUPD	Haemophilia A (Vivaud *et al.*, 1989)

maternal UPD resulting from the inheritance of both copies of a normal chromosome 15 from the mother (Mascari *et al.*, 1992). The reverse situation is observed for Angelman syndrome where paternal UPD is seen in 2% of cases (Nicholls, 1994). Cystic fibrosis (CF) has also occurred in two cases with UPD for chromosome 7 from mothers with CF mutations, whose fathers were found not to be CF carriers using haplotype analysis. Paternity was proven using DNA markers (Spence *et al.*, 1988; Voss *et al.*, 1989). Usually children with CF are of normal growth at birth, however, these 2 children with mUPD for chromosome 7 had both intrauterine and post-natal growth retardation. Patients with Beckwith–Wiedemann syndrome exhibit overgrowth and a predisposition to Wilm's and other embryonal tumours have been found in some cases to have paternal UPD for chromosome 11 (Henry *et al.*, 1991). Both these latter two examples have growth problems related to UPD and thereby once again implicate genomic imprinting effects that have a specific effect on growth.

Maternal uniparental disomy for chromosome 16 and IUGR

To try to assess the part that genomic imprinting or uniparental disomy plays in the aetiology of IUGR, chromosome specific DNA polymorphism analyses have been carried out in idiopathic IUGR families (Bennett *et al.*, 1992; Vaughan *et al.*, 1994).

Confined placental mosaicism, UPD and IUGR

Probably the most common mechanism for the occurrence of UPD is via 'trisomic rescue'. Trisomy 16 is the most common trisomy associated with both miscarriage and IUGR and therefore provides one of the first candidate chromosomes to study with respect to IUGR (Boue, Boue & Lazar, 1975; Kalousek, 1994). Cases where the trisomy 16 is complete with both the fetus and the placenta being affected account for 10% of the chromosomally abnormal miscarriages and have a phenotype of very early embryonic failure termed 'anembryonic' pregnancies. In other cases trisomy 16 is confined to the placenta with the fetus having an apparently normal karyotype and these have often been found to be IUGR pregnancies. Only recently has the DNA of the fetus been analysed to assess whether or not it shows UPD chromosome 16.

Human maternal UPD for chromosome 16 has been reported in 6 fetuses all associated with trisomy 16 confined placental mosaicism (Bennett *et al.*, 1992; Dworniczak *et al.*, 1992; Kalousek *et al.*, 1993).

Intrauterine growth retardation is the common feature. The contributions to fetal growth impairment between mUPD for chromosome 16 in the fetus and high levels of trisomic cells in the placenta cannot yet be ascertained (Kalousek *et al.*, 1993). Confined placental mosaicism occurs in about 1–2% of chorionic villus samples (Vejerslev & Mikkelsen, 1989), with trisomy 16 being common. The incidence of mUPD for chromosome 16 is unknown, but would expect to occur in about one-third of trisomic chromosome 16 confined placental mosaics (Hall, 1990).

IUGR and UPD for chromosome 16

Two of the severely growth related fetuses found to have mUPD for chromosome 16 and trisomy 16 placental mosaicism both had an unfavourable outcome (See Fig. 1) (Vaughan *et al.*, 1994). Antenatally the first case was complicated by an unexplained raised maternal serum alphafeto-protein concentration, preterm rupture of the membranes and growth retardation detectable at 21 weeks gestation. The other had an unexplained raised maternal serum human chorionic gonadotrophin level, a 2 vessel cord on ultrasound and cessation of growth at 25 weeks. At postmortem both babies had an imperforate anus. Fetal mUPD 16 may explain the poor outcome that occurs in some cases of confined placental mosaicism for chromosome 16 and could also be associated with specific fetal abnormalities such as imperforate anus.

Genomic imprinting, IUGR and the insulin-like growth factors

One approach to assess the contribution of genomic imprinting and UPD to IUGR is to analyse those chromosomes that have a high incidence of confined placental mosaicism. A second approach is to study those chromosomes that contain growth factors that would obviously be important in normal fetal growth. A third selection procedure is to then screen those growth factors against the published imprinted mouse map and to see if the mouse homologues are in an established imprinted region. Some members of the insulin-like growth factor family fit both these criteria.

Insulin-like Growth Factors I and II (IGF-I, IGF-II) are polypeptides with molecular weights of approximately 7 kD and homology to proinsulin. IGF-I mediates the actions of growth hormone on cartilage and bone formation whilst IGF-II may have a role in fetal growth and

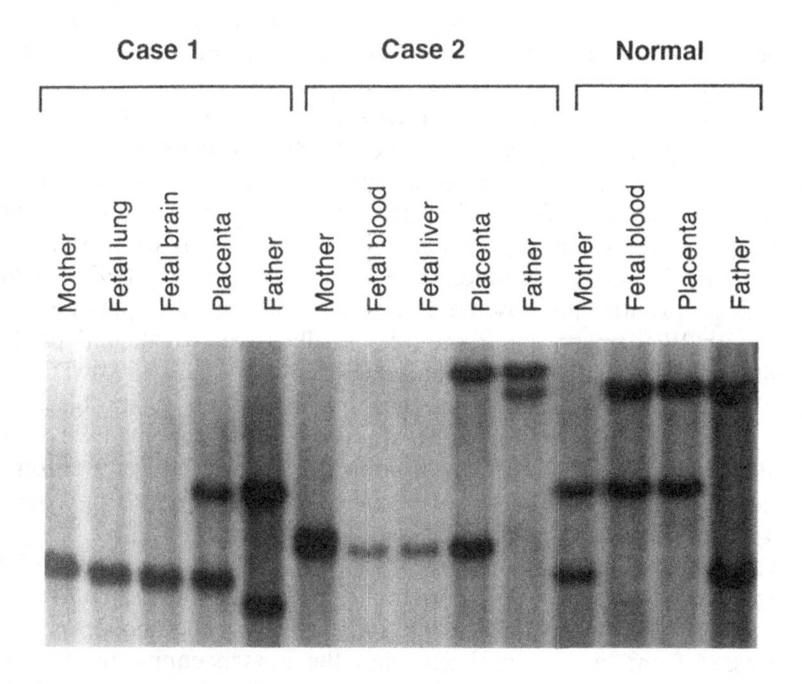

Fig. 1. A Southern blot of two IUGR families and one control family showing maternal uniparental disomy for chromosome 16. Fetal and parental DNA from both mUPD 16 cases and a normal control are digested with Taq I restriction enzyme and hybridized to the chromosome 16 VNTR probe p79-2-23 (16q22–24). In Case 1, the mother was homozygous and the father heterozygous for p79-2-23. All the fetal tissues show only maternal alleles. The mosaic placenta shows 2 maternal and 1 paternal allele. In Case 2, both parents were heterozygous. Fetal blood and tissue shows only the 2 alleles from the mother, whilst the mosaic placenta has 2 from mother and 1 from father. The normal control shows heterozygous parents with the fetus and placenta inheriting 1 allele from each parent (Vaughan *et al.*, 1994, reprinted by permission of John Wiley & Sons, Ltd. 12/8/94).

development (Binkert *et al.*, 1989). Their effects are modulated by a family of at least six binding proteins (IGFBP1 to 6) of which IGFBP1 and 3 appear to be the most important in post-natal life (Holly & Martin, 1994). There are also two receptors IGFIR and IGFIIR. IGFIR resembles the insulin receptor and binds both IGF-I and IGF-II but has a 20 times higher affinity for IGF-I. IGFIIR only binds IGF-II but is possibly not as important as IGFIR in the modulation and control of IGF-II.

In mice the *Igf1* gene is on chromosome 10 which is not considered to be a region displaying imprinting effects (Beechey & Cattanach, 1994). The mouse *Igf2* gene is on chromosome 7 (distal) and it has been shown that both *Igf2* in the mouse and its equivalent in the human IGF2 (on chromosome 11p15.5) are imprinted and only expressed from the paternally derived chromosome (Giannoukakis *et al.*, 1993). In the human pUPD for chromosome 11 has been diagnosed in patients with Beckwith–Wiedemann syndrome who exhibit overgrowth and predisposition to Wilm's and other embryonal tumours (Hall, 1990). Disruption of normal IGF2 gene expression has been seen in fibroblast cell lines and tongue tissue from Beckwith–Wiedemann patients (Weksberg *et al.*, 1993). Hypothetically, in an individual with mUPD for chromosome 11, the IGF2 gene would be inactive and may well predispose to IUGR. This has yet to be described.

The *Igf1* gene in the mouse and its equivalent in the human IGF1R, is expressed from both the paternal and maternal genes in fetal tissues (Villar & Pedersoen, 1993). The *Igf2r* gene in the mouse is only expressed from the maternal gene but the human equivalent IGF2R, is expressed from both alleles in fetal tissue (Kalscheuer *et al.*, 1993). Unlike *Igf2* and IGF2 the imprinting of the *Igf2* receptor in mouse and man do not correlate.

IGF binding proteins 1 and 3 have been localized to chromosome 7p12–14 in the human which is syntenic to an imprinted region in the mouse on chromosome 11 (Beechey & Cattanach, 1994).

Evidence from early and pilot studies are now beginning to implicate abnormalities in the IGF axis in IUGR. Studying paired samples of maternal and cord blood serum in IUGR births and matched controls shows that the serum IGF-I concentrations are significantly lower in the IUGR babies whilst their mothers have concentrations indistinguishable from the control mothers (Vaughan, Jones, & Preece personal communication). It is thus possible to develop a line of reasoning that idiopathic IUGR may, in part, depend on imprinting processes that act either through the IGF axis or in other ways at present undefined.

Summary

From the information detailed above it is clear that genomic imprinting as a phenomenon has a potentially large part to play in the control of growth in the fetus. It has been shown to be important in the mouse with several genes that have been found to be imprinted. Three of these are from the IGF family, a family of growth factors, receptors and their binding proteins that are known to be important in growth and development. Other growth factors as yet not analysed may well also be imprinted and need further study with respect to their role in the aetiology of IUGR.

The imprinting effects observed in the mouse are not always mimicked in man so simple studies expecting homology must be treated with caution. However the studies in the mouse do pinpoint chromosome areas of interest in the human. There are now many examples of uniparental disomy in man that exhibit an imprinting related phenotype. Many of these have growth disorders. As uniparental disomy only gives information for the whole chromosome the actual genes involved in most cases remain as yet uncharacterized.

Trisomic rescue is probably the most common mechanism by which uniparental disomy can occur. Trisomy for chromosome 16 is one of the most common trisomies found in miscarriage and trisomy 16 confined placental mosaicism has been associated with IUGR. More data on the genotype of the fetus with confined placental mosaicism is needed to assess the part that maternal UPD for chormosome 16 has to play in IUGR. Early onset severe intrauterine growth retardation was present in both our cases and was also the consistent feature in all four previously reported cases (Kalousek *et al.*, 1993; Dworniczak *et al.*, 1992). Whether it is the mosaic placenta, the fetal maternal UPD or both synergistically that cause impaired fetal growth remains unknown. Detection of a placental mosaic for chromosome 16 should always prompt further investigation of fetal tissues using DNA polymorphisms to trace parental origin as this may seriously affect the pregnancy outcome.

Acknowledgements

The research into the role of UPD in IUGR by these authors is supported by grant B2/92 from Wellbeing (formerly Birthright).

References

Antonarakis, S.E., Blouin, J-L., Maher, J., Avramopoulos, D., Thomas, G. & Conover Talbot, C. Jr. (1993). Maternal uniparental disomy for human chromosome 14, due to loss of a chromosome 14 from somatic cells with t(13;14) trisomy 14. *American Journal of Human Genetics*, **52**, 1145–52.

Barlow, D.P. (1994). Imprinting: a gamete's point of view. *Trends in Genetics*, **10**, 194–9.

Beechey, C.V. & Cattanach, B.M. (1994). Genetic imprinting map update. *Mouse Genome*, **92**, 108–110.

Bennett, P.R., Vaughan, J., Henderson, D.J., Loughna, S. & Moore, G.E. (1992). Association between confined placental trisomy, fetal uniparental disomy and intrauterine growth retardation. *Lancet*, **340**, (Nov 21), 1284–5.

Binkert, C., Landwehr, J., Mary, J.L., Schwander, J. & Heinrich, G. (1989). Cloning, sequence analysis and expression of cDNA encoding insulin-like growth factor binding protein. *EMBO Journal*, **8**, 2497–502.

Boue, J., Boue, A. & Lazar, P. (1975). Retrospective and prospective epidemiological studies of 1500 karyotyped spontaneous human abortions. *Teratology*, **12**, 11–86.

Brzustowicz, L.M., Allitto, B.A., Matseoane, R., Theve, R., Michaud, L., Chatkupt, S., Sugarman, E., Penchaszadeh, G.K., Suslak, L., Koenigsberger, M.R., Gilliam, T.C. & Handelin, B.L. (1994). Paternal isodiomy for chromosome 5 in a child with spinal muscular atrophy. *American Journal of Human Genetics*, **54**, 482–8.

Cattanach, B.M. & Beechey, C. (1990). Autosomal and X-chromosome imprinting. *Development*, (Suppl.) **103**, 63–72.

Cattanach, B.M. & Kirk, K. (1985). Differential activity of maternally and paternally derived chromosome regions in mice. *Nature*, **315**, 496–8.

Crouse, H. (1960). The controlling element in sex chromosome behaviour in *Sciara*. *Genetics*, **45**, 1425–43.

Dobson, P.C., Abell, D.A. & Beischer, N.A. (1981). Mortality and morbidity of fetal growth retardation. *Australian and New Zealand Journal of Obstetrics and Gynaecology*, **21**, 69–72.

Dworniczak, B., Koppers, B., Kurlemann, G., Holzgreve, W., Horst, J. & Miny, P. (1992). Uniparental disomy with normal phenotype. *Lancet*, **340**, 1284–5.

Engel, E. & DeLozier-Blanchet, C.D. (1991). Uniparental disomy, isodisomy, and imprinting: probable effects in man and strategies for their detection. *American Journal of Medical Genetics*, **40**, 432–9.

Fancourt, R., Campbell, S., Harvey, D. & Norman, A.P. (1976). Follow-up study of small-for-dates babies. *British Medical Journal*, **1**, 1435–7.

Giannoukakis, N., Deal, C., Paquette, J., Goodyer, C.G. & Polychroakos, C. (1993). Parental genomic imprinting of the human IGF2 gene. *Nature Genetics*, **4**, 98–101.

Hall, J.G. (1990). How imprinting is relevant to human disease. *Development Supplement*, **103**, 141–8.

Hall, J.G. (1992). Genomic imprinting and its clinical implications. *New England Journal of Medicine*, **326**, 827–9.

Henderson, D.H., Sherman, L.S., Loughna, S.C., Bennett, P.R. & Moore, G.E. (1994). The association between uniparental disomy for chromosome 21 and early embryonic failure. *Human Molecular Genetics*, **3**, 1373–6.

Henry, I., Bonaiti-Pellie, C., Chehensse, V., Beldjord, C., Schwartz, C., Uterman, G. & Junien, C. (1991). Uniparental disomy in a genetic cancer-predisposing syndrome. *Nature*, **351**, 665–7.

Holly, J.M.P. & Martin, J.L. (1994). Insulin-like growth factor binding proteins: a review of methodological aspects of their purification, analysis and regulation. *Growth Regulation*, **4**, 20–30.

Kalousek, D.K. (1994). Current Topic: confined placental mosaicism and intrauterine fetal development. *Placenta*, **15**, 219–30.

Kalousek, D.K., Langlois, S., Barrett, I., Yam, I., Wilson, D.R., Howard-Peebles, P.N., Johnson, M.P. and Giorgiutti, E. (1993). Uniparental disomy for chromosome 16 in humans. *American Journal of Human Genetics*, **52**, 8–16.

Kalscheuer, V.M., Mariman, E.C., Schepens, M.T., Rehder, H. & Ropers, H-H. (1993). The insulin-like growth factor type-2 receptor gene is imprinted in the mouse but not in humans. *Nature Genetics*, **5**, 74–8.

Kitchen, W.H. (1968). The relationship between birthweight and gestational age in an Australian hospital population. *Australian Paediatric Journal*, **4**, 29–37.

Lawler, S.D. (1984). Genetic studies on hydatidiform moles. *Advances in Experimental Medical Biology*, **176**, 147–61.

Li, E., Beard, C. & Jaenisch, R. (1993). Role of DNA methylation in genomic imprinting. *Nature*, **366**, 362–5.

MacFadden, D.E. & Kalousek, D.K. (1991). Two different phenotypes of fetuses with chromosomal triploidy: correlation with parental origin of the extra set. *American Journal of Medical Genetics*, **38**, 535–8.

Martin, R.H., Rademaker, A.W., Hildebrande, K., Long-Simpson, L., Pederson, P. & Yamamoto, J. (1987). Variation in the frequency and type of sperm chromosomal abnormalities among normal men. *Human Genetics*, **77**, 108–44.

Mascari, M.J., Gottlieb, W., Rogan, P.K., Butler, M.G., Waller, D.A., Armour, J.A.L., Jeffreys, A.J., Ladda, R.L. & Nicholls, R.D. (1992). The frequency of uniparental disomy in Prader–Willi syndrome: implications for molecular diagnosis. *New England Journal of Medicine*, **326**, 1599–607.

McGrath, J. & Solter, D. (1983). Nuclear transplantation in the mouse embryo by microsurgery and cell fusion. *Science*, **220**, 1300–2.

McGrath, J. & Solter, D. (1984). Development of mouse embryogenesis requires both the maternal and paternal genomes. *Cell*, **37**, 179–83.

Nicholls, R.D. (1994). New insights reveal complex mechanisms involved in genomic imprinting. *American Journal of Human Genetics*, **54**, 733–40.

Spence, J.E., Perciaccante, R.G., Creig, G.M., Willard, H.F., Ledbetter, D.H., Hejtmanuk, J.F., Pollack, M.S., O'Brien, W.E. & Beaudet, A.L. (1988). Uniparental disomy as a mechanism for human genetic disease. *American Journal of Human Genetics*, **42**, 217–26.

Surani, M.A.H., Barton, S.C. & Norris, M. (1984). Development of reconstituted mouse eggs suggests imprinting of the genome during gametogenesis. *Nature*, **308**, 548–50.

Vaughan, J., Ali, Z., Bower, S., Bennett, P.R., Chard, T. & Moore, G.E. (1994). Human maternal uniparental disomy for chromosome 16 and fetal development. *Prenatal Diagnosis*, **8**, 751–6.

Vejerslev, L.O. & Mikkelsen, M. (1989). The European collaborative study on mosaicism in chorionic villus sampling: data from 1986 to 1987. *Prenatal Diagnosis*, **9**, 575–88.

Villar, A.J. & Pedersen, R.A. (1993). Spatially restricted imprinting of mouse chromosome 7. *Molecular Reproduction and Development*, **37**, 247–55.

Vivaud, D., Vivaud, M., Plassa, F., Gazengel, C., Noel, B. & Goossens, M. (1989). Father-to-son transmission of Hemophilia A due to uniparental disomy. *American Journal of Human Genetics*, **45**, A226.

Voss, R., Ben-Simm, E., Avital, A., Zlotogna, Y., Dagan, J., Godfrey, S., Tikochinski, Y. & Hillel, J. (1989). Isodisomy of Chromosome 7 in a patient with cystic fibrosis: could uniparental disomy be common in humans? *American Journal of Human Genetics*, **45**, 373–80.

Weksberg, R., Shen, D.R., Fei, Y.L., Song, Q.L. & Squire, J. (1993). Disruption of insulin-like growth factor in Beckwith–Wiedemann syndrome. *Nature Genetics*, **5**, 143–9.

Wramsby, H., Fredga, K. & Leidholm, P. (1987). Chromosome analysis of human oocytes recovered from preovulatory follicles in stimulated cycles. *New England Journal of Medicine*, **316**, 121–4.

Index

Note: page numbers in *italics* refer to figures and tables